choosing and keeping pigs

choosing and keeping pigs

a complete practical guide

Linda McDonald-Brown

NOTE

This book is intended to give general information only. The publisher, author and distributor expressly disclaim all liability to any person arising directly or indirectly from the use of, or any errors or omissions in, the information in this book. The adoption and application of the information in this book is at the reader's discretion and is their sole responsibility.

An Hachette UK Company
www.hachette.co.uk

First published in Great Britain in 2009 by
Gaia Books, a division of Octopus Publishing Group Ltd
2–4 Heron Quays, London E14 4JP
www.octopusbooks.co.uk

ISBN: 978-1-85675-311-1

A CIP catalogue record for this book is available from the British Library.

Printed in Slovenia

10 9 8 7 6 5 4 3 2 1

Contents

Introduction 6

Getting Started 18

Caring for Pigs 46

Pests and Diseases 66

Breeding 78

Processing 96

Pigs on Show 106

Pig Breeds 116

Questions and Answers 192

Glossary 202

Index 203

Acknowledgements 208

Introduction

Pigs are sociable, intelligent animals that will give you many hours of entertainment and pleasure throughout their lives. This chapter explores the history of pig-keeping and looks at how to select a breed. One thing is for sure: your life will never be the same again!

Why keep pigs?

Pigs can give you years of enjoyment, just as much as any cat or dog, as long as you give them the space and care they need. You don't need a reason for becoming a pig owner – pigs are simply a pleasure to keep!

A passion for pigs

Ask any smallholder why they keep pigs, and the chances are you will be with them for the next couple of hours – after which you will leave none the wiser. You will, however, know everything there is to know about their pigs. As you walk through the pig pens listening to their owner regaling you with stories of their favourite pigs over the years, it will occur to you that, whatever the reason, their pigs have become an all-consuming passion.

Rearing your own meat

Most smallholders start out by buying and raising two weaners (newly weaned piglets) just for their freezer. Sitting down to home-produced pork is not only an extremely tasty experience, but you also have the satisfaction of knowing exactly what went into your meal. Many smallholders experiment with different sausage recipes or try their hand at curing bacon, which they then either use for their own consumption or sell locally.

Keeping a couple of weaners for the freezer doesn't take a huge amount of work, and the cost of getting them to porker weight is reasonable when you consider the amount of meat you get in return. However, pigs should not be seen as a 'get-rich-quick scheme', for feed costs can fluctuate wildly. Whether you plan to keep ten or a hundred pigs, a budget needs to be set and a close eye kept on it.

Today pedigree weaners are in demand. If you are planning to breed in the future, selling newly weaned pigs is a good way of maximizing profits faster than breeding weaners to fatten them on and sell for meat. However, the initial outlay of breeding pedigree weaners can be considerable. Buying in-pig pedigree gilts (female pigs that haven't yet produced a litter) can be expensive, depending on the breed. However, if your gilt has a large healthy litter that is sold at eight weeks for a good price, you should get back the money you paid for her (and more).

Pigs soon learn when feeding time is due and will wait impatiently for you at the gate until dinner appears.

Many pig owners progress to selling the meat they produce on a more professional but small-scale footing through cooperatives, local shops or farmers' markets. If done properly, keeping a close eye on costs, this can earn smallholders a decent profit, because the demand for local free-range meat is growing rapidly.

Other reasons for keeping pigs

If breeding for meat doesn't appeal to you, but you would still like to find a use for your pigs, buy a couple of older pigs and put them to work on any rough land or woodland that needs clearing. Pigs are renowned for digging up ground, and in no time at all will clear nettles, thistles, weeds and other unwelcome growth. If you are planning to plant young trees in woodland, put pigs in first to clear any bracken and briar.

Alternatively, you may wish to keep pigs for the pleasure of showing them and reaping the rewards of your careful husbandry and pedigree stock (see Pigs on Show, pages 106–115). Or you may want pigs to help you become self-sufficient in food, alongside growing your own vegetables; or simply to bring a touch of the farmyard to your suburban lifestyle, whether you keep pigs as pets or for their meat. Whatever your reasons, your plans should not be set in stone. What often starts out as a bit of fun, with minimal time and effort required, frequently turns into an all-consuming passion.

Docile and intelligent, Kune Kunes love the company of humans and make perfect pets.

Pig-keeping regulations

The regulations concerning pig husbandry can change from time to time, so it is a good idea to keep in contact with the appropriate government department or local supervisory body for up-to-date advice. Traditional breed clubs can also be a useful source of information.

How to use this book

This book is aimed at anyone who wishes to keep a pig for the first time, particularly novice smallholders who need guidance on how to care for their animals and get the most from them. Getting Started (pages 18–45) offers advice on essential equipment and preparing for your pigs' arrival. Caring for Pigs (pages 46–65) describes the daily, weekly, monthly and annual tasks, explaining how to feed, water and handle your pigs. Pests and Diseases (pages 66–77) discusses common pig ailments and how to treat them.

Breeding (pages 78–95) looks at getting your pig in-pig, the birth and caring for piglets. Processing (pages 96–105) explains what happens to pigs when they are processed as meat. Pigs on Show (pages 106–115) covers showing your pig and the etiquette and ethics involved. Pig Breeds (pages 116–191) is a directory of all the main pedigree breeds, detailing each breed's appearance, character and care needs. Finally, the questions and answers section on pages 192–201 deals with frequently asked questions about choosing and keeping pigs.

The history of the pig

Pigs are one of the most versatile creatures around, and over the centuries they have provided us with food and clothing and products such as buttons, artists' brushes and fertilizer; they have also played an important part in the medical industry.

The first domesticated pig breeds were descendants of the wild boar, which can still be seen roaming wild in some countries.

Domesticating pigs

The pig has been around for thousands of years and is believed to have been one of the first animals to be domesticated. Early breeds of the domesticated pig were descendants of the wild boar (*Sus scrofa*). Pig remains have been found at Neolithic sites, including one in Wiltshire in England, and cave paintings thousands of years old, such as those at Altamira in Spain, show wild pigs and humans together.

Domestic pigs historically have been raised in one of two ways: either by keeping them confined to a pen (the method that is familiar to us today) or by the more natural method of allowing them to forage over a wide area (a method that is still popular in some countries). Pigs are omnivores and, left to forage for themselves, will survive by eating almost anything, including roots, fruit, reptiles and carrion.

Early practices in England

Pigs were an important source of food for the Anglo-Saxons and were looked after by swineherds who took them to pannage. This is a legal term for the practice of turning pigs out into the forest at certain times of year to forage for foods such as acorns, beech nuts, roots and berries. Pannage came to play an important role in woodland ecology and is still practised today in parts of southern England.

Up to the Middle Ages, pigs were probably the main source of meat for humans and were kept in most villages. However, the widespread practice of letting them out to forage meant that they were often viewed as pests. To discourage unsupervised foraging, stray

Before the Second World War, many rural families kept a pig to supply them with pork and bacon.

Pigs in folklore

It is said that English sailors and fishermen used to regard pigs as very unlucky – so much so that they would not say the word 'pig' at sea or even allow pork or bacon on board. It was, however, fine to say 'grunter' or 'porker'. If they met a pig on the way to their ship, they would not sail that day.

In the county of Sussex, pigs are respected for their independent spirit and are even associated with the county's informal motto: 'We wun't be druv' (driven).

And, according to legend, the Gloucester Old Spot, a breed that originated many years ago in the orchards of Gloucestershire, gained its black spots from apples falling from the tree and bruising its skin.

pigs were killed or the pigs were impounded and the owner charged for their return.

Traditionally, most pigs (apart from breeding stock) were killed at the beginning of winter to provide the villagers with much-needed meat to keep them alive during the cold season. Those pigs that were not killed often wintered out in the forests, with no extra food.

The popularity of pig-keeping slowly declined after the Middle Ages, due to the growth in sheep-breeding and restrictions on pannage. A new type of pig was now kept – the cottagers' pig, less hardy than the forest pig. During the 18th century most cottagers kept at least one pig, feeding it on leftovers and waste produce from the vegetable garden. They would rarely breed from their pig, instead buying it at weaning and keeping it until it was ready for killing a few months later.

Changing fashions

During the 18th and 19th centuries in Britain, important improvements were made in the selection and breeding of pigs. Prick-eared pigs of Chinese and Siamese bloodlines were introduced to improve existing breeds, establishing the pig breeds that we are familiar with today. The Berkshire, for example, displays the prick ears and slightly snub noses characteristic of Asian pigs. In contrast, the Tamworth, unlike other breeds, has not been influenced by Asian breeds and remains the closest relative of the wild boar.

Fashion and the market have affected the fortunes of traditional British breeds over the years, with many breeds coming close to extinction and others (such as the Lincolnshire Curly Coat and the Dorset Gold Tip) being lost for ever. Until the 1930s traditional breeds were produced commercially, but after the Second World War the market changed and the specialist pork pigs and bacon pigs declined. Demand increased for the Large White and for the Essex and Wessex breeds (now known as the British Saddleback, after an amalgamation took place in 1967). Pigs such as the Berkshire and the Middle White, which had been renowned for their pork, declined to dangerously low levels and it is only in recent years that they have started to recover. The Middle White, at one time so popular with the capital's butchers that it was known as the 'London Porker', is now making a slow comeback as more and more restaurants put it on the menu. The popularity of the Large White continued to grow and in 1952 it overtook all other pigs to become one of the most important breeds in commercial pork production.

Following the recommendations of the Howitt report in 1955, the government asked the British farming industry to focus on three breeds – the Large White, Landrace and Welsh. By the 1970s most traditional British breeds were all but lost, but thanks to a handful of determined breeders and the Rare Breeds Survival Trust they are now regaining ground. As more people become aware of the provenance of their food, they are turning to traditional breeds raised naturally. Farmers' markets and the internet are increasing the availability of pork products from traditional breeds.

Lard, meat and bacon pigs

In general there are three types of domestic pig: the lard pig, such as the Mangalitza (see pages 156–159); the pork (meat) pig, such as the Berkshire (see pages 132–133); and the bacon pig, such as the Tamworth (see pages 138–141). Lard pigs are not as popular as they were: when the lard market declined after the Second World War, numbers of Mangalitza and other lard pigs fell considerably. However, due to the development of a market for cured meat, it and other lard pigs are now growing in number.

Pigs are not just needed for meat and lard. Pig by-products play a part in the manufacturing of many everyday items. Pigs are also important in the medical world: their heart valves have been successfully transplanted into humans for a number of years, and pig pancreas glands are a source of the hormone insulin, which is used to treat diabetes.

Today, pigs are once again living and foraging in some of the world's forests, as they did thousands of years ago. Some Scottish estates, for example, have reintroduced pig herds to help in the management and conservation of their woodland. Such initiatives can only be beneficial for the future of pigs, which must surely continue to be very much a part of our lives.

The Large White is still one of the most popular breeds for commercial pig farmers.

The Gloucester Old Spot is a traditional breed that has once again become popular for its succulent meat.

Choosing the right breed

When selecting a breed of pig, you will need to take into account your own circumstances and environment, as well as the pigs' character and needs, to ensure both your enjoyment and the welfare of the pigs.

Pure or cross-breed?

Choosing the ideal breed of pig for you and your family can seem daunting to a novice. Do you go for cross-breeds or pure pedigree types? Do you opt for large or small, coloured or plain pigs? It is important to consider first what you are planning to do with your animals. Are you keeping them as pets, for the freezer or are you hoping to breed from them? Do you need them to root or would you prefer pigs that graze?

Whatever your reasons for buying a pig, it is important that you find out what you are getting. Cross-breeds certainly have their place, especially in the meat trade, where cross-breeding is undertaken to improve the quality, taste and texture of meat. However, even an

The Tamworth is a mischievous pig that loves to root – it is not for the beginner.

old hand can never say for sure how an eight-week-old cross-bred weaner will turn out, for it will have the temperament and characteristics of two different breeds. A Tamworth (see pages 138–141) crossed with a wild boar (see pages 162–165), for instance, is definitely not a pig for a novice; it can be aggressive and an amazing escape artist. A Gloucester Old Spot (see pages 126–127) crossed with a Saddleback (see pages 128–131) will probably have a more placid temperament, but if you are planning on keeping pigs for meat, you will need to consider if this cross is suitable for the type of cuts you have in mind.

Buy a pure-bred weaner and you can say with more or less certainty how it will turn out. At the same time, you will be helping to keep these beautiful breeds going. Most breeds have their own club and it is worth joining for access to a list of local breeders and a regular newsletter that contains a wealth of information.

Space

Take a good look at your own circumstances and what space is available to you. A large, mischievous pig like the Tamworth is better suited to a free-range environment such as woods and will not be happy in an urban or semi-urban garden, no matter how large it is. However, the Kune Kune (see pages 184–187), a smaller and friendlier pig, will probably adapt quite happily to a more confined area – a space at least 10 x 10 metres (30 x 30 feet) would be suitable for a pig of this size, as long as all its welfare requirements are met.

Rooter or grazer?

You also need to take into account that pigs tend to root and can turn a lovely green pasture into a mudbath within weeks. However, some breeds root more than others. In general, pigs with long snouts (such as the Tamworth, see pages 138–141) tend to root, while pigs with snub noses (such as the Middle White, see pages 144–145) graze and do less damage. This is certainly something to bear in mind if you are planning to keep a couple of weaners in your garden for the freezer. On the other hand, if you are planning to use your pigs to turn over a scrubby piece of land, you definitely need to buy pigs that root.

In general, pigs with lop ears (ears that cover their eyes), such as the Large Black (see pages 134–137),

Pig courses

Spending a day on a pig course is an excellent way of getting up close and personal with the different breeds. Choose a course that is 'hands-on' and not only will it give you the confidence to handle your pigs when they arrive, but it will enable you to experience the different pig temperaments at close quarters. Often the sheer size of these animals can be a real eye-opener for a novice pig-keeper.

Kune Kunes are small and non-aggressive pigs that could be kept in a large garden.

Oxford Sandy and Black (see pages 124–125) and Gloucester Old Spot, tend to be quieter and more easy-going than the prick-eared breeds, and are often recommended for beginners.

Type of meat

If your pigs will be solely for eating, decide exactly what sort of meat you would like. If you want to fill your freezer with joints, chops or sausages, it is worth looking at the Berkshire (see pages 132–133), which is world-famous for its succulent taste – and much prized by the Japanese. Known as the 'Ladies' Pig' because of its dainty size and attractive features, it is an easy pig to keep – perfect for novices who only have access to a small plot of land. For bacon, the Tamworth has to be top of the traditional breed list and is worth considering once you have gained experience with other breeds.

Agricultural shows

Once you have researched the different breeds, make a list of the ones that you feel would meet your needs (and whose needs would be met by you). Having narrowed down the search, it is time to go one step further and meet the breeds that you have shortlisted.

During the summer, agricultural shows are widely held, many of which have pig classes. For the novice, this is an excellent opportunity to meet the different breeds all at once and see exactly what you would be taking on. At the top shows you can watch the *crème de la crème* of the pig world and see how they handle. Viewing the various breeds in top condition in a show environment should help you decide if a particular breed is for you. At the end of the day, you need to like the appearance of the pig, for you will be looking at it every day!

Once the classes are over, breeders like nothing more than talking about their pigs, and you may find there is an opportunity for you to go inside the pens with a breeder and get to know your prospective breed at close quarters.

Local breeders

An alternative to agricultural shows is to visit local breeders. As long as you give plenty of warning, many breeders will be happy to show round a potential future customer. Be honest with the breeder as to why you wish to keep pigs. If you are going to keep them purely as pets, there is no point going to see a breeder about a Landrace (see pages 120–123), which is first and foremost a commercial meat pig. You should be looking at something smaller, such as a Vietnamese Potbelly (see pages 188–189) or a Kune Kune (see pages 184–187). If you are in any doubt about which type to get, the breeder should be able to help you decide how suitable the breed is for your requirements.

The easiest way to make friends with a pig is through its stomach – fruit and vegetables are much-appreciated treats.

Keeping traditional pigs is the best way of helping to ensure these breeds survive into the future.

Breeding pigs

If you wish to breed from your pig, it is a good idea to research the local market before investing in stock. If, for example, there appears to be a saturation of Large Black breeders in your area, a sensible option would be to go for a different breed, such as the Oxford Sandy and Black, for you might need a market for up to 12 piglets in any one litter. The situation may change over time, but at least you will be able to get a foot in the door.

It is important that you realistically assess the amount of land you have available for the pigs. Breeding pigs require greater amounts of land. Should you not be able to sell the weaners as quickly as you had hoped, you will have to keep them separate from the sow. So buying a pig with the intention of breeding in the future should not be undertaken lightly; if you only have a small back garden at your disposal, then getting your beloved gilt in-pig should be put on the back burner until you have more land.

Obviously, if you are planning to breed, you should start off with pure-pedigree registered pigs. All the traditional breeds would be in danger of extinction if another reversal of their popularity occurred, as it did in the 1960s. By buying, breeding or even just eating pure-breds, you are helping to conserve pigs for future generations. In the event of an outbreak of disease such as foot-and-mouth, your traditional pigs may be exempt from slaughter if they are pedigree and registered.

At the end of the day, there is nothing more satisfying than looking out of your window and seeing pure-bred traditional breeds snuffling around the trees and knowing that you have played your part in helping to conserve them.

Getting Started

Once you've made the decision to start keeping pigs, you need to prepare for their arrival. This chapter will guide you through the processes of buying the right equipment, setting up a home for your pigs, acquiring stock and record-keeping.

Equipment

You need not spend the earth when you are starting to keep pigs, but you will need to invest in the essential start-up equipment. Buy the best-quality equipment you can afford as items designed specifically for pigs will prove better value in the long term.

Budgeting

It is worth spending some of your budget on some essential start-up equipment that will in the long run make your life easier and your pigs more comfortable and possibly also save you money. Much of the other equipment listed later in the chapter can be bought as and when it is needed, so you can spread the cost over a period of time, rather than buying everything at once.

When sourcing equipment, research as many different suppliers as possible, because prices may differ considerably from company to company. It is also worth looking at the possibility of buying direct from the manufacturer, even if they are based in a different country. Sometimes buying equipment in this way (especially smaller items) can work out cheaper than buying from a local company. However, when buying over the internet, try to ensure that the quality of the item in question is as good as the website says it is.

Many smallholders, when they buy their first weaners – especially if they are working to a tight budget or don't plan to keep pigs in the long term – 'make do' with whatever equipment they have lying around. For a temporary or emergency measure it is certainly possible to use cheap alternatives to some pieces of equipment. For example, you can build a shelter from straw bales instead of purchasing a specially built ark; you can use buckets for holding water instead of fitting costly automatic drinkers; and, in dry weather, you can

A pig ark needs to be built from weatherproof material and should be strong enough to withstand the natural destructive tendencies of your pigs.

scatter feed on the ground instead of placing it in troughs. However, if done on a permanent basis, these practices could turn out to be expensive, wasteful or labour-intensive. Straw shelters won't last as long as purpose-built arks and have to be continually rebuilt; buckets, no matter how secure they seem, will end up being tipped over and, more often than not, broken.

Essential start-up equipment

Suitable housing, drinkers, a trough, bedding, rat-proof feed storage and a slapboard and pig stick are all vital items that are best bought before your pigs arrive.

Housing

Ideally you should house your pigs in a purpose-built ark. It should be sturdy and made of strong, pressure-treated timber and exterior-grade fittings. For ease of movement, it should come with skids (ski-like structures) on the bottom or a loading bar on top for lifting it. Try to steer clear of fixed-floor arks: not only are they harder to clean out, but if you suffer the misfortune of having a sick pig or piglet in the ark, it is virtually impossible to get to it easily. Instead, buy an ark with a removable floor, which not only enables you to give the ark a thorough cleaning, but also provides easier access to sick pigs should you need to administer injections or medicines.

Drinkers

It is possible to use temporarily a water bucket held securely in a supporting base such as a car or tractor tyre, but ideally you should fit nipple drinkers or automatic drinkers. Fix these so that the pipes leading to them cannot be damaged by inquisitive pigs. Both types of drinker should be checked daily for problems, and automatic drinkers should be cleaned at the same time. Should it not be possible to fit either of the above, then troughs can be used, but they are easily knocked over so they should be wedged against something solid and immovable.

Pigs need lots of water, especially in hot weather, so drinking troughs should ideally be positioned next to a water supply.

If you do not have a suitable place for attaching a drinker, and the enclosure is too far away from a water supply to be able to carry water to a trough, it may be necessary to buy a water butt or some other container that will capture and store rain. If the container is open to the elements, consider placing some mesh over the top to discourage birds from drinking from it and potentially contaminating the water with their faeces. If you are using this method of providing water for your pigs, be aware that, should you experience a long dry spell, the container will dry up and you will need to have plans in place for bringing water to the pigs by some other means.

Troughs

If the pigs are being fed large nuts (see page 54), many owners scatter the nuts on the ground when it is dry, because foraging for them keeps the pigs occupied for longer. However, feeding in this way is not ideal if you are giving the pigs smaller nuts or mash, or if the ground is muddy. Not only will you be throwing your money away, because most of the food will be trodden into the ground and wasted, but your pigs will not get the correct amount of food. In the end, scattering food on the ground will lead to loss of condition or slower weight gain in porkers, so it is far better to feed your animals from a trough.

One of the most popular troughs used by pig-keepers is the round 'Mexican hat' cast-iron feeder, which is split into sections and is ideal for smaller pigs and litters. The downside of this type of trough is that they are very heavy to move, quite expensive and hard to find new, although they can be picked up second-hand at quite reasonable prices at farm sales and auctions. For feeding larger pigs, long galvanized troughs in varying sizes are popular, as they are not too heavy and easily cleaned. Some companies make these shaped troughs in rubber.

Bedding

Providing your pig with appropriate bedding is a necessity. Apart from the obvious comfort benefits, it reduces leg injuries and the possibility of infections caused by the pig dunging and urinating in the pig house. Whichever type of bedding you use, it should not be skimped on and should be put down thickly to ensure warmth and protection, especially in winter.

Straw is the bedding most widely used among pig owners and is readily available. However, it is not that easy to store, requiring a lot of space unless you buy it in very small quantities. Check the bales carefully for any dust or mould, which could make your pigs ill. Straw is ideal for piling around the entrance to the ark to minimize the amount of mud dragged into it. However, straw should be used with caution when bedding down newly born piglets, as they tend to bury themselves underneath, thereby running the risk of being lain on by the sow.

An alternative to straw is wood shavings, although these are not as warm and insulating and can be expensive. Wood shavings come in different grades, and it is worth going for the highest-quality grade, which tends to be less dusty (but costs more). Shavings

Easy-to-clean galvanized troughs can be used for providing both food and water.

The 'Mexican hat' feeder is suitable for smaller breeds and litters but, being made of cast-iron, can be extremely heavy.

are often popular with breeders, who use them to bed down sows and young piglets.

Shredded paper can also be used as bedding, but it is hard to get hold of, expensive and, although it looks fresh and clean when first put down, can get soggy quite quickly and give the impression that the bed is dirtier than it is.

Feed and equipment storage

Ideally food should be kept in a building, such as a shed, that is dry and rat-proof and separate from the area in which your pigs are kept. Any food should be emptied from the sacks in which it arrives into specially made animal-feed containers as soon as possible, to prevent spillage and keep the food fresh. If your budget doesn't run to buying specially made containers, it is perfectly acceptable to use a chest freezer that is no longer in working order.

It is useful to have a few buckets spare for carrying feed and water. Choose ones with strong handles, otherwise you could find the contents all over the floor. Spare buckets, your medicine cabinet and (if the shed is large enough to hold it) bedding can also be kept in your food shed.

Straw should be checked for dust or mould before use and laid down thickly on the floor of the ark.

Strong buckets are always useful for transporting feed and water, so keep some handy around the pen.

Feed containers need to keep the feed fresh and protect it from the weather as well as from rodents and other animals.

The slapboard and stick are used together to encourage a pig in the right direction.

Slapboards and pig sticks

Slapboards (also known as pigboards) are a must when moving pigs about or loading them into a trailer – for your own protection, as well as for moving the pigs more easily. They are used to direct a pig along the required route, by placing the boards at the side of the pig's head to encourage it to turn in the opposite direction. Sizes differ, but the boards are usually about 60 cm (2 feet) square and made from plastic, fibreglass or 8-mm (⅜-inch) plyboard. All types have a hand-hole at the top. Some manufacturers make extra-large slapboards that need two people to hold them, and these are useful if you are moving a few pigs at the same time.

Also required is a pig stick about 60 cm (2 feet) long. Used in conjunction with the slapboard, the stick encourages forward movement. Handling sticks can be purchased from specialist shops, but smallholders usually make one themselves or use something like a shepherd's crook.

Recently a combination of slapboard and stick, known as a 'paddle', has become available. Lightweight and with plastic beads inside that sound like food in a bucket, it looks just like a paddle – hence its name.

Identification equipment

Smallholders who plan just to keep a couple of weaners for the freezer don't need to go to the expense of buying lots of identification equipment. In many countries, even if the pigs you buy are pure-bred and registered with the relevant breed society, they will come already identified, so all that is required for the abattoir is for you to put a metal tag and/or slapmark with your herd number on the animal. Many breeders will actually metal-tag your pigs for you before you take them home, which will save you the cost of having to buy your own ID applicator.

However, if you are planning on breeding and registering the offspring, you will have to buy the equipment required to identify that particular breed (see Record-keeping on pages 44–45). Whenever you are identifying a pig, ensure that your instruments are as clean as possible. It is a good idea to spray the ear with an antiseptic as a matter of routine once you have finished marking it.

Temporary marking paste

This is useful if you need to identify a pig for a specific reason or are moving pigs under a year old from farm to farm. Different-coloured marking paste can be bought from most specialist pig companies and normally comes in a push-up container. It should be legible until the pig reaches its destination at the very least, but ideally should remain on the pig for about seven days.

Ear notchers

These are used for identifying coloured breeds, such as the Large Black and Saddleback. In the UK there are two different breed systems of notching, one for the British Saddleback and a different one for other coloured breeds. Ear notchers come in different sizes and can be bought from specialist companies or agricultural merchants.

Tattoo pliers and paste

Tattoo pliers with interchangeable characters are generally used for identifying all light-coloured pigs, such as the Middle White and Welsh. There is no preference about which colour of paste you should use, as long as the tattoo showing identification numbers and letters is clearly legible on the pig.

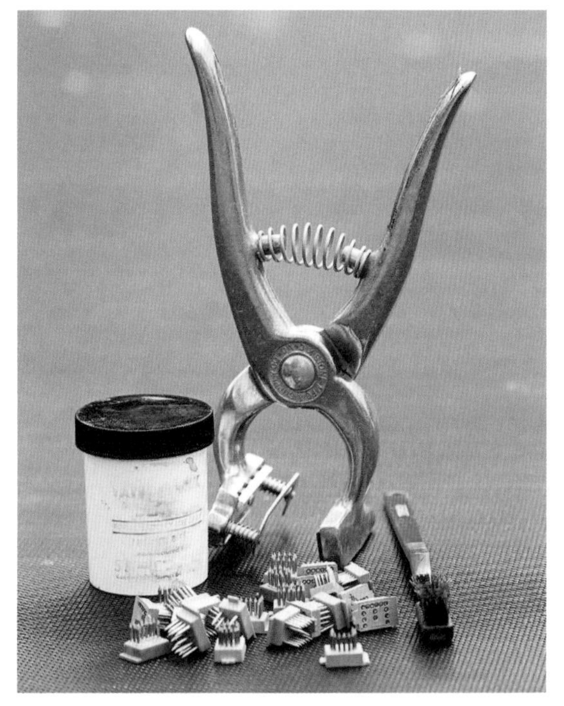

Tattooing equipment is used to stamp light-coloured breeds with identification numbers.

Slapper and paste

Slapmarking your pig with your herd number on its shoulder is essential if you are taking it to the abattoir. When collecting the carcass, you will be able to identify that it is your pig (and not someone else's) by checking the herd mark. Many smallholders slapmark on both shoulders, in case they receive back two separate halves. Slappers can be expensive to buy and are purchased from specialist companies. They come with either a fixed head (showing your designated herd number) or an interchangeable head (ideal if you have to slapmark pigs with different herd marks). The paste should be suitable for the job of slapmarking and should be fully synthetic.

A slapper with a head displaying your herd number is necessary for marking pigs for the abattoir.

Ear tags and applicators

Ear tags used for record-keeping are usually plastic and come in different shapes and colours to assist with record-keeping. Tags used for slaughter must be metal to withstand the scalding process. All tags should be printed with the country of origin and your herd number. When buying ear tags, make sure you are getting the correct applicator for the tag.

Carrying box

Buy a lightweight plastic box for storing and transporting all your identifying equipment. This can be picked up in a saddlery, DIY store or kitchen shop.

General cleaning equipment

There is no getting away from it: handling pigs usually means getting dirty and smelly. Choose equipment that will stand up to the job in mind, as there is nothing worse than trying to clean out a pig house with implements more suited to the garden.

Clothes

To save on clothing it is worth keeping a set of 'pig clothing'. A good-quality pair of overalls is a must, as are rubber boots. Choose the most expensive rubber boots you can afford, as cheaper versions split and won't last five minutes. In the winter you will need a warm, windproof coat, a warm hat and gloves – fingerless ones are usually best, as they keep your hands warm, but still enable you to do fiddly tasks. It is also a good idea to buy a few pairs of waterproof over-trousers. These will see quite a lot of use, so it is better to buy a few cheap pairs than one expensive pair that could get ripped the first time you wear them.

Wheelbarrow

Choose a wheelbarrow that is light and easy to push, with rubber handles. Electric wheelbarrows are easy to move, even when full of wet straw, but may be hard to manoeuvre around corners and narrow spaces.

Rubber boots and overalls are essential for venturing into your pigs' muddy environment.

Forks, shovels and brooms

Choose brooms that have hard bristles; a fork with long, curved prongs; and a shovel that is lightweight, with a slight curved edge to enable you to scoop up dirt easily. It is also worth buying a long-handled scooper to pick up dung from the pig paddock.

Hosepipe

Although not essential, a hosepipe will make your life easier when you are cleaning out arks, trailers and runs, and when filling troughs and wallows. Make sure it is put away tidily after use, to prevent inquisitive pigs chewing it.

Disinfectants and mats

Every pig owner should have to hand disinfectant, as this plays an important part in keeping infectious disease at bay. Ideally, you should buy both powder and liquid disinfectants, plus a specialist disinfectant to use in a footbath. Before buying disinfectants, take advice on which will suit your needs, then follow manufacturer's instructions exactly. (For more about bio-security, see pages 62–65.)

Useful miscellaneous items

Other useful items include a pocket knife, record book and torch. Don't invest in your own trailer until you are sure that pig-keeping is for you.

Sharp pocket knife

A sharp knife is a must for cutting open straw bales. Make sure that the knife you buy is retractable. It is a good idea to attach an empty bobbin or something similar to your knife with brightly coloured bailer twine, to stop you from losing it.

Nose rings and applicator

Nowadays pigs are rarely ringed, as they were in the past to prevent them from rooting up ground. The general opinion is that if you keep pigs, you must expect them to root up the ground and curbing that natural instinct is cruel. Pig rings come in two types: large steel rings that are inserted through the septum of the nose (most effective because of their size) and smaller copper rings that go through the side of the nostril. Both types of ring are inserted using a specialist ring applicator.

Record book

A notebook is useful for recording litters born and other matters when you are out with your pigs. Choose one that is small enough to fit in your pocket and has a washable cover. You might want to attach a pencil to it by means of a piece of string.

Torch

A torch is essential when checking pigs at night. Choose one that you can wear on a head strap in order to leave your hands free for carrying items and opening gates.

Horse brush

Pigs like nothing better than a good scratch. If they have dry skin, give them a firm brush with a small horse brush, prior to rubbing them in pig or baby oil. Horse brushes can be purchased at agricultural merchants and saddleries.

Thick plastic sheeting

When trying to keep a sick pig warm in its ark, it may be necessary to close off as much cold air as possible. If the ark has a door, you simply shut it. If there is no door, you can use thick plastic sheeting cut into vertical strips and nailed to the top of the entrance to keep the ark draught-free. Plastic sheeting can be found at most builders' merchants and DIY stores. Alternatively, feed sacks or hessian can do the job equally well.

Pig weighing band

Although it is hard to find these specialist weighing bands, it is worth spending time looking for one. They work like a tape measure and give an accurate reading, in both kilograms and pounds – a godsend to novice pig-keepers who can't yet judge weight by eye.

Trailer

Ideally you should have your own trailer, although you may not want to go to the expense of buying one until you know that pigs are definitely for you and that you intend to keep them for many years. You don't necessarily have to have a specialist pig trailer; many people borrow or hire one, or use a horse or cattle trailer for transporting pigs. Check before you transport the pigs that your car is able to pull the weight of the trailer plus the animals.

Medical equipment

Small herds of outdoor pigs rarely experience health problems. However, it is a good idea to keep a small selection of medicines to hand, to deal with any minor problems that may arise. Remember to check the 'use by' date frequently and to throw out any that have passed it, even if they have not been opened. Not all of the items listed here will be needed if you only have a couple of pigs that are going to slaughter after a few weeks, so choose those that you think most relevant to you.

- **Suncream** This is essential if you have light-coloured pigs or piglets because they are prone to sunburn (see page 71). Choose a high SP-factor suncream and rub it on all the light-coloured areas, especially around the tops of the ears. If your pig is light-coloured all over, it may be worth keeping it in during the day and turning it out at night, or providing a wallow – covering themselves in mud is the best protection from the sun that pigs can have. Should your pigs suffer from sunburn, apply calamine lotion to cool the skin.
- **Pig oil and baby oil** These are ideal for use on pigs with dry skin, and are popular with breeders for their show stock.
- **Vinegar** This can be used for cooling down pigs if they are suffering from heatstroke (see page 70); a small container of vinegar should always be kept in the medicine cabinet.
- **Wound powder and antiseptic spray** Keep a general antiseptic spray or powder for cuts and small wounds. This can be purchased from veterinarians, agricultural merchants or over the internet.
- **Wormers and other parasite products** Keep in stock anthelmintic preparations for the control of worms (see page 71) and external parasites, which can be purchased from vets.
- **Needles and syringes** Speak to your vet about the size of needle suitable for your size of pig. Keep needles and syringes for administering wormers and antibiotics.
- **Thermometer** This is essential for taking the temperature of a sick pig.
- **General-purpose antibiotic** It is useful to keep a general-purpose antibiotic for emergencies. Take advice from your vet on which one is suitable. Do not use it past its 'use by' date.
- **Cotton wool** A small roll of cotton wool should be kept for cleaning wounds and eyes.
- **Rubber gloves** Wearing rubber gloves when cleaning wounds, administering injections or identifying your piglets is good practice.
- **Twitch** Although not strictly a piece of medical equipment, a twitch is used to restrain pigs during a veterinary examination or when it is necessary to administer injections. Placed around the snout, it consists of a loop attached to a handle. Avoid wire twitches as they can cause damage to the tissues of the snout if incorrectly used. A rope twitch is kinder and works in exactly the same way.
- **Small cupboard** Try to keep all your medical equipment and products in one place. A cupboard or small refrigerator that is no longer working is ideal. Make sure that children cannot get to the contents by placing a small lock on the door.

Fencing

Badly erected fencing will be no barrier to pigs, so it is important that a perimeter fence is strong and well constructed, to deter even the most determined of pigs. There are two main types of fencing: post-and-rail and electric fencing.

Escape artists

Unlike other farm animals, which may happily stay confined within a fence that is sagging slightly or in need of attention, pigs will give any weaknesses (no matter how small or insignificant) their undivided attention until the desired result has been achieved: escape. They are notorious for escaping from their enclosures and will do so, given the chance, by digging, pushing or even jumping their way out. And once your pigs are out, you run the risk of them causing an accident or creating damage by rooting on neighbouring land or even attacking people or other animals.

Fencing the perimeter correctly is therefore of the utmost importance when preparing the area in which you will be keeping your pigs. Ideally pigs should be kept behind solid walls, although for most owners this is not possible. It is vital that the fencing you use is of good quality and strongly constructed and, once up, that it is checked regularly and repaired quickly when necessary. Many years ago it was not unknown for sows with piglets to be tethered using a shoulder harness: the piglets would stay close and the sow would not be able to stray, and they would then be moved to fresh ground as and when required. Today, this method is no longer used.

Post-and-rail fencing should be made from good-quality stock netting placed as close to the ground as possible.

You can strengthen post-and-rail fencing with a top rail. The posts should be at least 20 cm (8 inches) in circumference.

Fencing considerations

Try to fence as large an area as possible, for pigs kept within a spacious site tend to be more contented to stay within its boundaries than those kept in too small an area. Keep no more than six pigs per 0.5 hectares (1 acre) and try to rotate them regularly around different parts of the enclosure.

Be conscious of the type of land you are planning to fence. Some land, such as peat, is very soft, which can cause problems with the stability of the straining posts. Measures such as driving in extra-long posts should be taken to ensure good durability. Driving in a thin metal pole until you hit solid ground gives you a good indication of exactly how long the straining posts should be. Ascertain the ideal length of the posts before ordering them.

Think carefully about where to site the gates. Ideally there should be two gates to every enclosure: a smaller one for everyday access and a larger one to enable you to take a trailer into the area. Gates should be hung so that they open both inwards and outwards, if possible. To prevent the pigs from putting their snouts underneath the gate and literally lifting it off its hinges, one hinge should always be placed upside down. There is no clear preference for steel or wooden gates, although wooden ones may be susceptible to chewing by the pigs. Whichever type of gate you choose, buy the best you can afford.

Safety warning

If you have any electric fencing situated near a public footpath or in a place that the public will visit, make sure that you put up clearly visible warning notices.

When planning your fencing, it is a good idea to take into account any expansion or breeding plans that you might have for the future. If you are intending to place your pens close together, consider making provision for walkways between them. Paths laid with woodchip or concrete between enclosures are not only useful for keeping pigs well away from each other (especially important if you have more than one boar), but will enable you to feed and water the animals easily, without actually coming into contact with them.

Post-and-rail fencing

The most popular choice for enclosing pigs is sturdy post-and-rail fencing. High-tensile pig netting or stock netting is then used as a filler and is stretched between large, round posts that are at least 20 cm (8 inches) in circumference, with the bottom of the fence as close to the ground as possible. If you prefer, instead of a rail along the top, a strand of barbed wire or electric fencing could be used, to deter all but the most determined pigs from escaping. To strengthen and support the fence even further, straining wire can be placed along the bottom, middle and top of the fence.

Whichever method of fencing you choose, remember that it will only be effective if it is done properly. So if you have never put up fencing before, now is not a good time to practise.

Electric fencing

Electric fencing is ideal for use as temporary fencing to partition off parts of the enclosure to enable you to rotate the pigs. To keep in adult pigs successfully, you need an energizer. This can be run using either a mains energizer unit, which is the best option as it is less likely than the portable unit to go wrong, or a portable energizer unit with a battery; both types need to be checked daily to ensure that they are still charging and that the wire has not been shorted out by clods of earth or long grass. It is not unusual to find the odd dead frog or toad stuck to the wire.

If you plan to use this type of fencing, it is worth investing in a tool for measuring the strength of electricity being carried along the wire. However, electric fencing will only be effective if the pigs are used to it. So before erecting it and expecting the pigs to stay within its boundaries, teach them to respect it.

One way of doing this is to put your pigs in an enclosed building, such as a barn or stable, and leave them in there for about a week with a strand of electric fencing placed just inside the door about 23 cm (9 inches) off the ground. The pigs soon learn that every time they go near the wire, they get a shock. Placing the wire in front of a closed door also means the pigs are unable to barge forwards through the electric fencing, which is what usually happens at first in an open environment. Once pigs have learned to respect electric fencing, it is possible to keep whole herds behind just two strands of wire a few inches off the ground.

Electric fencing can also be effective in discouraging pigs from digging underneath pig netting. Use two strands, the first placed approximately 15 cm (6 inches) off the ground and the second placed 23 cm (9 inches) above that, to ensure that all but the smallest piglets are stopped in their tracks, once their noses catch the wire. If you are not keen on using electric fencing for this purpose, then a strand of barbed wire can be just as effective.

Electric fencing is powered either from the mains or by a portable battery unit. Either way, the fencing should be checked daily.

Housing

Some form of shelter must be provided for pigs at all times of year. It should be large enough to enable your pigs to turn around comfortably and lie down and ideally sufficiently sturdy to withstand their destructive tendencies.

Shelter considerations

In winter pigs need shelter from the wind and rain, and in summer they need protection from the sun. A few trees in the corner of the paddock won't suffice, and it should not be assumed that pigs kept in woods will get sufficient shelter from the trees, no matter how dense the wood is.

When choosing or constructing a shelter, bear in mind that pigs – and boars in particular – are very destructive, and unless the shelter is made to a high standard and built from quality materials, it will not last. In the long run, buying the best shelter you can afford will be the most cost-effective. A shelter must also be large enough to accommodate comfortably the number of pigs that are using it. It is better to have too much space than too little – an extra bale of straw can always be added to fill up the space and keep your pigs warm.

Shelters that are positioned in the wrong place can have a detrimental effect on the welfare of your pigs, so careful thought needs to be given to their siting. If you are using an existing building, you obviously cannot change its position, but you should check that the building is draught-proof as far as possible, has good ventilation and has nothing in it that could injure the pig, such as exposed nails in the walls. If the building is old, check that the walls are sound: bricks that have come loose will attract the attention of your inquisitive pigs, and eventually the strength of the wall could be undermined. Ideally the building should be positioned as close as possible to an outdoor area to give the pigs some freedom.

Ark sizes

Below are sample ark sizes for housing different numbers of pigs:

1.8 x 1.2 metres (6 x 4 feet)
two Kune Kune pigs

2.4 x 1.8 metres (8 x 6 feet)
one sow and litter/up to two dry sows/six to eight porkers

2.4 x 2.4 metres (8 x 8 feet)
up to four dry sows/10–15 porkers

2.4 x 3 metres (8 x 10 feet)
up to six dry sows/15–20 porkers

2.4 x 4.2 metres (8 x 12 feet)
up to eight dry sows/20–25 porkers

Some of the more popular types of shelter are described below (for information on farrowing arks and sheds, see page 82).

Arks

By far the most common shelters for pigs are traditional pig arks with curved roofs. These sorts of pig shelter are favoured by commercial farmers of outdoor pigs and they are usually made of metal or wood and galvanized steel.

Choose an ark with ventilation and (if possible) removable floors, which are easier to clean than fixed floors. The floors should be made of solid rough-sawn timber rather than plywood, which has a tendency to become slippery when wet or muddy. Arks are also manufactured in plastic, but these can become brittle and crack. To aid movement, arks should come with a loading bar or skids.

To counteract fluctuations in temperature, you can add insulation to the ark, ensuring that the temperature is kept reasonably constant throughout summer and winter. Insulation will help porkers gain weight, as the food they eat will be used in gaining weight rather than in keeping warm. Arks without floors are not as warm as floored arks and they should only be used on well-drained soil.

Choose a sheltered position when placing the ark in a field. Face the entrance away from the prevailing wind and rain and, if your land is prone to bogginess, position the ark on a high point. If your land is very exposed, make sure that the ark has some form of shade. Placing it under trees is ideal, as the leaves will provide shade over the ark in the summer. If this is not possible and your ark does not have insulation, you need to make alternative arrangements to keep the interior cool in the summer, such as placing pieces of carpet or straw bales over the top of the ark to maintain an even temperature.

The most popular choice of housing is the curve-roofed pig ark, made from metal or wood.

Brick pigsties were once a popular form of pig shelter, but can be small and hard to clean.

Traditional pigsties

During the 19th and early 20th centuries, pigsties were a familiar sight on farms and at many of the large country estates. Traditional breeds were often kept on the big estates to feed the workers and the owner's family. The pigs were housed in low, brick buildings, which often had a brick pen attached. Although they are aesthetically pleasing to look at, these original pigsties are very difficult to clean and move about in, because they usually have small entrances and the space inside is limited.

If you want to build a row of pigsties in the old-fashioned style, keep height and space in mind as you plan their construction. Try to position the shelters so that the entrance faces south, to get as much sun as possible. Each pigsty should be at least 1.8 metres (6 feet) square. The floor should be constructed of ridged concrete to help with grip, and it should be insulated, to keep the pigs as warm as possible. Drainage is an essential feature and should be incorporated in some form, both from the pigsties and the outside pen. A window that can be opened and shut is essential for ventilation. Fitting doors is a matter of preference, but they are useful for when you need to keep your pigs in their sty.

Barns

Barns are ideal for housing a number of pigs together, such as store pigs (pigs for fattening). They are usually light and airy, but can be draughty. Ideally you should be able to get in with a tractor and a scraper to clean them out, because mucking out using just a wheelbarrow and a fork will be too time-consuming. Barns are often popular with commercial farmers as they enable store pigs to be kept warm and thus gain maximum weight in a short period of time. If you do keep pigs in this way, you will need to provide some sort of entertainment for them, such as a football or a length of rope or chain suspended from the roof with a tree branch attached to the end.

Home-made straw shelters

Straw shelters are ideal for erecting in an emergency or as a short-team measure. However, they are not suitable as permanent features, although if they are substantial enough, they can last a surprisingly long time before they need replacing. The structure will last longer if you build a framework first.

Decide how large you want your shelter to be and lay down one layer of straw bales. Place a post on the inside of each corner. Determine where you wish the entrance to be, then position four posts, one at each corner of the bales at the entrance. Finish infilling with bales up to the required height. Once you are happy with the structure, line it with substantial plywood on the inside by nailing this to the posts. Make sure there are no nails protruding. The plywood prevents the pigs from pulling at the straw, which will eventually weaken the shelter. Galvanized corrugated sheets, overlapped to prevent the rain from penetrating, should then be placed across the length of the shelter. If it is a large shelter, you may wish to add a wall of extra posts and bales to support the roof. Try to give the roof an overhang, to protect the straw sides from the worst of the weather. For insulation, place a layer of straw bales on top of the sheeting.

To discourage infestations of lice or mites, dust the shelter regularly with an approved powder. Should you need to keep the pigs inside, you can simply place a couple of straw bales across the entrance.

A straw shelter is a useful temporary measure that will last longer if built over a framework.

Acquiring stock

Choosing a healthy pig that catches the eye, and that is free (as far as possible) from fault, is essential when selecting stock. Buying pigs that are registered with the relevant breed society can help you choose good-quality stock.

Choosing your pigs

Once you have studied and selected your preferred breed and have decided whether to buy breeding stock or start off with a couple of weaners for the freezer, it is time to start looking around for a breeder.

For first time pig-keepers, an ideal time to buy weaners is in the early or late spring. By that time the best of the weather is still to come and looking after them won't seem such a chore (no matter how much you love your pigs, caring for them is always much harder in the winter).

Joining the relevant breed society will enable you to access its list of registered breeders. Beware, though: just because they are members of a breed society doesn't necessarily mean they sell quality stock, and for this reason you must always visit the breeder and spend some time with him or her.

If you are having difficulty sourcing a local breeder, try looking in specialist magazines and agricultural newspapers or on the internet. Often breeders who are showing their pigs at agricultural shows will have an information board detailing the stock they have for sale. Avoid buying stock at an auction or livestock market. Pigs sold in these places are usually of poor quality, or are pigs that the owners have been unable to sell privately for some reason. If no one else wants them, why should you?

Never buy pigs over the telephone without seeing them first, unless you have brought from that particular breeder before and know their stock (and they know your requirements). Good breeders care about where their weaners go and whether they will be looked after properly. Be wary if breeders seem unwilling to offer help and support once the pigs have left their property

When choosing your weaners, try to pick the liveliest and healthiest-looking animals you can.

– it may mean that the weaners have something wrong with them. To be on the safe side, go elsewhere to purchase your pigs.

If a breeder doesn't have weaners available when you require them, it is worth asking if you can be put on a waiting list. Often well-known breeders have sold their pigs before they are even born, so it is only possible to buy by being put on their waiting list. This might mean quite a long wait.

Inspecting the pigs

When making an appointment to visit, try to fit in with the breeder. It's best not to make appointments around lunchtime or in the early evening; and if you have to cancel, give the breeder as much notice as possible – don't simply fail to turn up. It's common sense not to go looking at the pigs in your best clothing, and don't try to fit in a hurried visit between other appointments. You need to be prepared to spend time asking questions and getting a feel for the piglets before choosing. While you are there, make sure that you ask about worming vaccinations and feeding.

Try to look at a litter before any have been sold, so that you have a choice. Never buy if there is only one left – not only should you generally buy pigs in pairs (because they are sociable animals and can get depressed on their own), but the chances are that a lone piglet will be the runt, as all the other good piglets have been sold. Avoid buying any pig that you feel sorry for, because you could be setting yourself up for future problems.

Take note of the surroundings when you arrive, for the look of a property says a lot about the way things are run. It doesn't matter if the surroundings look old and tatty and the fence needs painting, but the pig pens in particular should appear organized and tidy. Fencing should be in a good state of repair, arks or shelters should have plenty of bedding and the pigs themselves should be happy to see the breeder. Don't be concerned if the pigs are walking around in mud up

Benefits of acquiring registered stock

- **Traceability** When buying a pure-bred, there is no guarantee that it is in fact a pure-bred (even if it looks like the breed in question) unless you can trace it back. Some crosses, such as the Duroc crossed with the Tamworth, can produce offspring that can look like either a pure Tamworth or a pure Duroc. To the novice, it will be impossible to tell. If the pig is not registered, then you could be buying anything.
- **Financial reward** Buying registered breeding stock means that you can ask more for the piglets when you come to sell them.
- **Meat-marketing schemes** Some countries have meat-marketing schemes that are open to producers of registered pedigree pigs and enable producers to sell their registered carcasses to butchers who have chosen to be part of the scheme. For instance, the Rare Breeds Survival Trust in the UK has a Traditional Breeds Meat Marketing Scheme.
- **Disease exemption** Having registered pedigree pigs may mean that you qualify for exemptions should there be a disease outbreak such as foot-and-mouth.
- **Conservation** By buying a pure-bred, you help to ensure the conservation of traditional breeds so that they do not die out.

to their stomachs – as you are about to find out, you very rarely have one without the other! Hopefully the breeder will already have separated the piglets into meat pigs and ones that are good enough to breed from. Obviously they will be priced accordingly, with breeding stock priced higher.

Take your time to watch the litter in their pen. When choosing pigs, it is always best to go with your first impression. A pig that has an air about it that says 'look at me' should always do well.

Whether buying for meat or to use for breeding, pigs should be well covered (healthy-looking, but not too thin or too fat), alert and curious to see you, and should have no signs of lameness or stiffness whatsoever as they walk around. Check their skin: it should be loose and healthy; be cautious if you see any redness or very dry skin, for it could be caused by lice or mange. Listen for coughing in the piglets, which could mean that the litter has worms. (See page 68 for further description of the healthy pig.)

A good temperament is vital if your pigs will be around children, so if possible take a look at both parents – if they have good temperaments, the chances are that the offspring will too. If you are buying with a view to breeding, temperament is even more important. Any nasty or nervous tendencies in either the boar or the sow will inevitably be passed on to some (if not all) of the offspring. If you find yourself with a particularly nasty boar or sow when it matures, you need to think very carefully about keeping it and you certainly shouldn't breed from it.

When buying pigs purely to raise for the freezer, correct conformation and colouring are not essential, as long as the pigs are healthy. However, buying your

Weaners should be neither too thin nor too fat, with healthy skin without lice or mange.

foundation stock (breeding stock) is a different matter. You should buy the best you can afford and as close to the breed standard as possible. Prior to buying your own pigs, it is worth viewing as many different pigs of your chosen breed as possible, to give you a good idea of how your ideal pig should look.

Buying breeding stock

By far the cheapest way of acquiring breeding stock is to buy a good-quality registered weaner at eight weeks. However, many people do not have the patience to wait for nine months before she can be put to the boar, and then another three months before she farrows. A much quicker (but more expensive) solution is to buy an in-pig gilt. This not only solves the problem of having to wait a while before you see the results, but is ideal for those who don't want to go to the trouble of having to find a boar or artificially inseminating her. Buying an in-pig gilt is also a good way to introduce beginner pig-keepers gently to the breeding process.

Breeding stock should be of the highest standard you can afford and certainly basic faults should be avoided. Lop-eared pigs such as the Large Black (see pages 134–137) should have lop ears, not pricked ones; Saddlebacks (see pages 128–131) should have clean saddles with no blue or black spots; Tamworths (see pages 138–141) should be a solid colour with no black hairs visible.

Whichever type or breed you choose, make sure that you are buying registered stock, and that the pigs have been earmarked in accordance with the breed society's requirements – if necessary, you should ask to see evidence of their registration.

Consider a pig's temperament as well as its physical health – a friendly, inquisitive pig makes a good choice.

Buying checklist

- ✔ Buy in early or late spring, if possible.
- ✔ Use the internet, specialist magazines and agricultural newspapers for contacts.
- ✔ Buy from a breeder where possible.
- ✔ Avoid buying at auction, at livestock markets or over the telephone.
- ✔ View litters before many piglets have been sold.
- ✔ Assess the state of the pigs' environment.
- ✔ Go with your initial impression when selecting your pigs.
- ✔ Check for good health and temperament.
- ✔ Check the pigs' earmarks, documentation and registration.

Preparing for your pigs' arrival

Treat your pigs like important visitors, anticipating all their needs. Their bed should be warm and dry, their food should be nutritious and easy to find and the perimeter of their shelter should be secure.

Minimizing stress

To a greater or lesser degree, all animals suffer anxiety when moving to a new environment. Preparation and organization are the key to minimizing these potentially stressful situations. Pigs in particular can suffer greatly from stress and, if the stress is prolonged, they can actually die of it. Illnesses such as pneumonia have been known to occur in pigs that have had to undergo a particularly long or stressful journey, and it is not unheard of for pigs participating in a show for the first time to drop dead from the stress of being taken away from their familiar surroundings and placed in a noisy environment.

As a rule of thumb, try to buy your pigs from no more than an hour's drive away, to keep their journey time to a minimum, and keep them in a confined area while bringing them home. If you are using a trailer, partition off a small part rather than allowing them the freedom of the whole trailer. Giving your pigs a confined area encourages them to relax, lie down and go to sleep, rather than investigate every corner of the trailer as you are moving along.

Forward planning

Start preparing for your pigs' arrival well in advance. Make sure that you have acquired all the necessary documentation for you to keep pigs legally. Some countries require you to obtain holding numbers and herd numbers (see page 44).

Plan to have everything in place to enable your pigs to settle down in their new home and acclimatize to an unfamiliar enclosure and ark, with no disturbances aside from feeding and watering. Apart from giving them a quick check when feeding and watering them, they should be left alone for the first two or three days.

If the pigs are going out into a large piece of woodland or field, it may be worth putting them in an enclosed building, such as a barn or stable, for the first week until they are happy and confident about approaching you. This eliminates the possibility of your pigs taking fright every time they see you and disappearing, which stresses them out and probably doesn't do much for you who has to chase them, either.

Assessing your site

Allow yourself a week from finalizing their new home to the arrival of your pigs, for any last-minute changes or repairs. Walk around your fencing at least twice before the pigs are let loose, checking for any holes or gaps that might have been missed the first time. The last thing you want are frightened pigs escaping into the neighbourhood.

Thoroughly check your enclosure for poisonous plants (see page 75). If you find any, pull up all of the plant (including the root) and burn it, as some plants such as ragwort are just as poisonous when lying on the ground dying as they are while growing. At the same time, make sure there are no dangerous objects sticking up through the ground that could harm an inquisitive pig.

Place the entrance of your pigs' ark away from the prevailing wind and in a quiet corner of the enclosure, away from children, dogs and other scary things, and throw in plenty of straw to keep the pigs warm. If your land is boggy or particularly wet, place the ark at the

Put lots of bedding inside the ark to make it comfortable for its new occupants.

highest point to prevent it getting wet and muddy. If your land is very exposed, place the ark in a sheltered spot, such as underneath trees or by a wall (see page 33). If your pigs are arriving in summer, supply them with a wallow. Some companies make galvanized steel wallows that can either be dug into the ground or sited above it. However, a simple hole, dug and filled with water, will suffice and is cheaper.

Food and water

Before you bring the pigs home, find out from the breeder the type of food they have been eating and buy in at least two weeks' supply, to enable you to change them over gradually to their new food. Whatever brand of food you are intending to feed your pigs, make sure that you source a supplier before you take delivery of the animals.

Place water and feed utensils (see pages 21–22) close to the ark so that nervous pigs don't have to wander very far to find nourishment. If possible, try to arrange for your pigs to arrive in the morning. This gives them ample time before dark to snuffle around, finding the boundaries and their feed and water supplies.

Finally, no matter how settled and contented your pigs look on their first day, resist the temptation to introduce them to all your acquaintances. Your neighbours won't mind waiting a few days before getting the camera out, and the pigs will be much happier if they have been allowed to settle in before being bombarded with visitors.

Garden pig-keeping

It is possible for a farm animal to be brought into an urban environment, as long as it is a suitable breed and its welfare needs are met. Keeping pigs in a semi-urban or even an urban setting is increasing in popularity.

Animal welfare

Up to the Second World War, most rural families kept at least one pig, ensuring a never-ending supply of pork – and it was not unusual to keep a pig in a back yard. Today we are far more conscious of animal welfare, and certain animal-welfare bodies would find it hard to accept a concrete yard as a suitable home for a couple of pigs. That said, the area you need in order to keep pigs is surprisingly small: just 10 x 10 metres (30 x 30 feet) will hold two weaners comfortably, until they are ready for the abattoir. However, the more land you have available for your pigs, the more natural an environment you will be able to give them.

Before becoming a suburban or urban pig-keeper, you need to appraise your house, garden and surrounding environment realistically. You may have the

Small-sized pigs, such as Berkshires, can be kept in a large garden.

largest garden in your street, but if the only way to it is through your house, then keeping pigs is not for you. It is vital that you have good access to your garden, preferably with a large gate through which you can back a trailer.

Urban considerations

Sourcing pig food in a town can be problematic, because, historically, agricultural merchants have always been based outside towns, so you may have to travel quite a way to find someone who can supply pig food. When you have found a supplier it is worth buying food in bulk, to save you money on fuel costs and to get discounted prices.

Finding a vet with good experience of pigs may also be difficult in an urban area, so it is worth researching this before you need one. Look in business directories for pig farms that you can ask for recommendations.

Neighbours are often a sticking point when it comes to keeping pigs in a suburban area, so invite them round before you get your pigs. Explain what pig-keeping will entail and, more importantly, whether it will affect them. To keep them enthusiastic, offer to give them some pork when the pigs go for slaughter. Above all, don't dismiss their concerns, or a previously friendly neighbour might turn out to be a thorn in your side.

You must be conscious at all times that you are bringing a farm animal into an urban area. The local environmental authorities will look closely at your plans to reassure themselves that your pigs are not going to suffer in any way, so be prepared for any questions they may ask of you.

If you go away two days out of every five, you need to have in place a reliable person to feed and water the pigs at least once a day. If they start to get hungry, they begin to look for ways of escaping and, once they learn a way out, you will have a hard job keeping them in.

You also need to think ahead, to when the time comes to have your pigs slaughtered, or you could find yourself with real problems trying to source a local abattoir. This might result in you keeping your pigs far longer than you anticipated and in them costing you more money.

Environment

Place the pigs and the enclosure as far away from property as possible. If you plan to keep pigs on an on-going basis, it might be worth planting some quick-growing trees to screen the pen.

Fence your pen properly (see pages 29–31); chicken netting simply won't keep pigs in. At the very least you need stock-proof fencing between sturdy wooden posts and, to make absolutely sure the pigs cannot stray, a strand of electric wire inside the fence.

Choose a shelter that blends in as far as possible with its surroundings and will stay looking decent for as long as possible. Straw can quickly become untidy, once the wind and rain get to it.

Choice of breed

If you live in an urban area, the type of breed you choose is critical for stress-free management. Some breeds, such as the Oxford Sandy and Black (see pages 124–125), grow too large for a garden; others, such as the Tamworth (see pages 138–141), are not only notorious escape artists, but are free-range pigs, more at home in woodland. Berkshires (see pages 132–133) tend to be popular with urban pig-keepers because they have a pleasant temperament and are reasonably small, although they take longer to reach a suitable slaughter weight. They are renowned for their pork, however, so it is worth the wait.

Whichever breed you go for, should your neighbours be unhappy at having pigs so close, you can reassure them that keeping pigs for the freezer is not a long-term project. They will be in the garden no longer than three or four months before they go to the abattoir – just don't tell them that you will be buying two more after that!

Record-keeping

Whether you have one pet pig or a herd of a hundred, keeping precise records is vital. These enable you to know exactly where your pigs come from and what medical treatment they have received, and to be capable of proving it, if necessary.

National requirements

Some records are solely for your own benefit, while others are for relevant government departments. If the latter are not kept up to date, you risk prosecution. Some countries require more official record-keeping than others, so check the legislation. Wherever you live, you should keep medical and breeding records as a matter of course, either on the computer or on paper.

Holding number/CPH number

In the UK, anyone who intends to keep pigs must, by law, obtain a holding number (CPH number), so that the government can trace all livestock in the event of a disease outbreak. A CPH number is a nine-digit number that represents the county and parish of the holding's location, as well as a number unique to you (e.g. 84/521/0045: county/parish/unique number).

CPH numbers should be applied for well in advance of the pigs arriving on your property. Contact either the Rural Payments Agency (RPA) if you live in England; or the Divisional Office of the Welsh Assembly; or the Scottish Executive Environment and Rural Affairs Department; or your local County Agriculture Office in Northern Ireland. To obtain the CPH number you will have to fill in and return a form. Occasionally the RPA requires a copy of a digital map of your property, which can be obtained from the Rural Land Registry (RLR).

Herd number

Once you have your CPH number and your pigs, you can apply for a herd number (again, you may be prosecuted if you don't have one). This number consists of two letters (identifying the county) and four digits (your own unique number) and is a means of identifying where the animal comes from. Contact your local Animal Health Office, which will require your holding number, name and address. You will normally receive your herd number within a couple of days or in some instances will be given it over the telephone.

Movement licence

Whenever you move a pig, a movement form must accompany it and details of all the movements must be

Identification

There are three ways of identifying your pig:

- **Ear tag** This must be stamped or printed (not handwritten), and must state your herd number and the country of origin. Tags that are used for slaughter must be metal to withstand the carcass processing; those used for general movement of animals can be plastic.
- **Tattoo** This is used as a means of identification on light-coloured pigs (see page 25).
- **Slapmark** This permanent ink mark detailing your herd mark is applied on both shoulders of the pig and must be legible (see page 25).

recorded in a movement book. The form requires the following information:

- Dates of travel (departure and arrival)
- Loading and unloading times
- Departure CPH
- Arrival CPH
- Name and address of premises of departure
- Name and address of premises of arrival
- Details and number of the animals being transported
- Haulier details

Send one copy of the form to your local Trading Standards Department within three days, and keep another copy for yourself. Movement records should be retained for six years after you stop keeping pigs.

Any pig that arrives on your property triggers a '21-day standstill', which means that no pig is allowed to leave your property during that time unless it is going straight to slaughter. If another pig arrives during that time, the 21-day standstill starts all over again.

Medical records

Medicines (including wormers and injections) must be recorded within 72 hours of administration and the records kept for at least three years. You will usually receive a medical record book when you are sent notification of your herd number. The following information is required:

- Details of animal treated and identification number
- Date treatment began and finished
- Name of the product and batch number
- Date withdrawal period ended
- Name of the person who administered the treatment

Breeding records

Although not required by law, diligent breeding records should be kept to enable you to know when a litter is due. The following information should be recorded:

A swift positive movement is required when applying the slapmark on your pig's shoulder.

- Date of seasons
- Names of parents
- Date the boar and gilt were put together
- Date of service (if known)
- Date litter is due
- Weaning date
- Details of any birth notification and herd registering

It is also helpful to record the number of piglets in a litter, to work out whether a particular sow is making or losing you money when you take into account the cost of feeding her and other incidentals such as straw and wormers. Some breeders sell on or cull a sow that regularly has small litters.

Breed association records

If you are a member of a breed association, you will be required to keep a record of all your registered pigs and their identification numbers. Every year you will be asked to update these records, which can be done online.

Caring for Pigs

Pigs are not difficult to look after, as long as you follow a few basic guidelines. This chapter tells you all you need to know about establishing a routine for your pigs, about feeding, watering and handling them and about ensuring that your hygiene measures are up to scratch.

Establishing a routine

Like most animals kept in captivity, pigs thrive on routine: not only will they be more settled if they are fed and watered at the same time each day, but they may well be easier to handle.

Benefits

By keeping to a routine, it is easier to spot or prevent problems that might otherwise have gone unnoticed. Just as importantly, certain tasks have to be carried out on a regular basis to help keep your pigs safe and well. And keeping to a routine also helps to build up a relationship between you and your pigs.

Daily tasks

Pigs need to be fed and given fresh water (if you are not using an automatic system) at least once a day, so this is an ideal time to carry out a mini health check. Start by watching them walk: are they stiff? Then, beginning at the head, run your hands and eyes over their body, legs and feet – first one side and then the other. By inspecting your pigs daily, you may be able to avert a potential health problem and save on vet's bills. Miss this daily inspection and you could find that a small cut, which at first would have required nothing more than a squirt of antiseptic to help it heal, has turned into an infected wound that needs to be treated with expensive antibiotics.

It is also a good idea to keep an eye on how much food each individual pig is consuming. Sometimes weaker or smaller pigs are unable to eat their full quota, due to larger or stronger pigs pushing them out, which can lead to potential health problems and loss of weight. If this is happening, move the pigs in question to another trough a little way off from the others.

Pigs soon learn when feeding times are due, waiting by the gate screeching and talking until the bucket appears and the nuts fall into their trough. Whether you feed them once or twice a day, try to ensure feeding always happens at the same time. Being late with the feeding or missing a feed could lead to bullying and fighting among your pigs, which in turn could lead to weight loss, injuries and stress.

Look inside your pigs' shelter to check that it is still clean and that no damage has occurred during the night through chewing.

Fencing should also be checked daily – any holes that have appeared underneath it need to be repaired with something more substantial than a few stones. If you have electric fencing running around the bottom of

At least once a day, you need to check that your electric fencing is still working properly.

Feeding time is also the perfect opportunity for a quick health check of your pigs.

your stock fencing, this needs to be cleared of any debris that may be lying across it. Pigs soon realize that if they push stones or sods of earth across the wire, it earths out and they can get on with digging their way out of the pen without fear of being 'zapped'. Electric fencing is only useful when it is working! It is a good idea to walk around your electric fencing at least twice a day.

Inspect automatic drinkers and clean them at least once a day, as mud and debris could restrict the flow of water.

During hot weather, you need to top up the wallow with water every day. Wallows are very important for your pigs' welfare, as they keep their temperature down and also help prevent sunburn on light-coloured pigs (see pages 70–71).

Carry out a weekly pen check to ensure any dangerous debris unearthed by the pigs is removed immediately.

Record-keeping (see pages 44–45) is tedious: all those forms to fill in, those registrations to send off to the relevant society, those moving forms to return within the allowed time limit... It is so easy to put them to one side for another day. Unfortunately, this not only leads to records that are out of date and to missed deadlines for registrations, but could also result in fines from authorities who do not receive the necessary records within the required time. It is therefore vital to set aside some time each day to fill in the forms and record books, so that everything is submitted on time and all your records are filed and up to date, should the authorities ever wish to see them.

Weekly tasks

At least once a week you need to carry out a thorough pen check. No matter which breed you own, pigs are notorious for digging things up: glass bottles, tin cans and rusty pieces of metal are often found poking through the soil by pig owners and must be removed. You should also look carefully at any new growth in the pen, especially around springtime. Do not assume that every plant growing there is a harmless weed; inspect it closely to reassure yourself that it isn't poisonous (see page 75). If it is, it needs to be pulled up in its entirety, removed from the pen and burned.

If you are lucky enough to have trees growing in the enclosure, keep a close eye on them. Pigs will damage trees if they continually root and snuffle around their base or rub against the bark, and it is worth paying particular attention to the roots. If the pigs are rotated around different enclosures on a regular basis, this tends not to be too much of a problem. However, if space is an issue and the pigs are kept on the same piece of land for any length of time, boredom can set in and the trees suffer.

If you find that the bark has been damaged or the roots have been exposed by snuffling pigs, the tree should be fenced off sooner rather than later to enable it to recover; otherwise it will die and the pigs will lose their summer shade.

Troughs and water utensils should be thoroughly cleaned two or three times a week, and more often if they get caked in mud. Ideally, move the utensil in question out of the pen and hose it down with hot water, using a scrubbing brush to get into the corners and leaving it to dry before putting it back in the pen.

Monthly tasks

Once a month the pigs' sleeping quarters need to be completely cleaned out. During the summer you may think this isn't necessary if the straw is reasonably clean and dry. However, by regularly cleaning out and disinfecting their sleeping area, you keep the possibility of infection down to a minimum.

The straw should be taken out and burned, and if possible the ark should be thoroughly scrubbed with hot water (if available), before being sprayed with disinfectant in accordance with the manufacturer's guidelines. If your pigs have got into the unfortunate habit of dunging in the ark, this cleaning process needs to be done more often (a removable floor will facilitate this unpleasant job). At the same time check for any sharp edges, which should be either hammered down or smoothed over.

If the shelter is built from straw bales, take out the bedding periodically and burn it, and replace the bales if they start to sag with water.

Worming – either by injection or powder in the feed – should be done every six months, so you should make a calendar note of when your pigs were wormed, so that you don't forget when they are due to be done again. Pigs arriving on your property from another holding should be routinely wormed, unless you are 100 per cent sure that they have been recently done. Once wormed, the pigs should be kept in a stable or barn for a few days to allow any worms or eggs to pass through, so as not to infect your land.

Routine checklist

Daily

- ✔ Fresh food and water
- ✔ Body inspection
- ✔ Individual food quota
- ✔ Shelter cleanliness
- ✔ Fencing
- ✔ Automatic drinkers
- ✔ Wallow levels (in summer)
- ✔ Record-keeping

Weekly

- ✔ Pen check
- ✔ Tree condition
- ✔ Utensil cleansing

Monthly

- ✔ Clean-out of sleeping quarters
- ✔ Following month's food delivery
- ✔ Worming (every six months)

Yearly

- ✔ Vaccinations

Arrange for next month's delivery to make sure you never run out of pig food. If the suppliers don't deliver, set a date every month to pick up the food yourself.

Yearly tasks

If you vaccinate your pigs, you must keep up with the annual boosters. Speak to your vet about when is the best time to vaccinate.

Feeding

Pigs love their food and are known for eating almost anything. Different breeds require different quantities of food to gain weight, but all pigs must be given the correct balance of vitamins, protein and fibre to remain healthy.

Differing requirements

There are no hard and fast rules. Each pig is different; some breeds reach the required weight faster and on less food than others. A Gloucester Old Spot (see pages 126–127), for instance, normally doesn't need the same amount of food as an Oxford Sandy and Black (see pages 124–125). The former tends to put on back fat, whereas the latter is usually longer and leaner and therefore needs that little bit more in order to reach the required weight. Show animals require that little bit extra again, as do pigs kept outside in winter and sows with a litter. Often correct feeding – and the way your pigs look – is all about developing an 'eye'.

Supply feed in a container if there is any danger that it will get lost in mud or straw on the floor.

Food basics

If you are lucky to have a choice of feed merchants locally, then you need to compare them on pricing and go with the cheapest – as long as the food they sell is of good quality and fits your requirements. Pig food makes up a huge percentage of outgoings and you should cut back in any way that you can.

It is worth investing in a silo. Some companies specialize in selling smaller silos for the smallholder; these are still expensive, but over the long term will save you money. The pig owner is in a much better position to negotiate a favourable price when buying in bulk. Even if you do not have access to a silo for loose food, you should always try to buy bags in bulk if possible. However, you will need sufficient space to store a large quantity of 25-kg (55-lb) bags.

Most people start keeping pigs because they care about the food they put into their mouths. Given the choice, the majority of pig owners would prefer to feed organically, but the price of organic feed means that this type of food is often out of the question for pig owners on a budget. If this is the case, try to buy food that does not contain Genetically Modified (GM) crops. This can still cost more than normal pig food, but is cheaper than organic.

In the wild, pigs eat little and often, very rarely going without food for too long. Feeding in this way is not possible when your pigs are confined, and the most usual method is to feed twice a day, in the morning and at night. Some owners feed once a day, but unless your pigs are on decent grass, this is not ideal as they will be left too long without food.

Piglets and weaners benefit from access to an ad lib feeder from which they can take small amounts of feed at a time.

If possible, piglets and weaners should be fed ad lib rather than at set times. Young pigs can suffer from scours (diarrhoea, see page 72), if suddenly given feed after being left too long without it. A small hopper full of food left in the creep (the piglets' nest area, see page 83) will enable the piglets to eat little and often. Breeders feeding by this method often find that the piglets gain weight faster than those fed twice a day. Creep food should be the best you can buy and should be sweet-smelling and free from dust. This food normally contains a large amount of cooked cereals, milk proteins, lactose and fishmeal and is more expensive than other compound foods, but it is worth paying the extra to ensure that your piglets have the best start in life.

Ideally pigs should also have access to grass. All pigs graze, but some breeds such as the Tamworth (see pages 138–141) have longer snouts and are therefore better equipped for rooting rather than grazing; whereas a pig like the Middle White (see pages 144–145) has a shorter nose and tends to graze rather than root. However, if the grass runs out – no matter which breed you have – it will root. Pigs such as the Kune Kune (see pages 184–187) can actually do quite well on grass alone in the summer, provided the grass is of decent quality and there is enough of it. In any case you should avoid giving smaller breeds more than 16 per cent protein in their diet, and preferably less. Try to buy pig food that meets this requirement for the smaller breeds, although it is much more difficult to source than ordinary pig compound food.

Compound food

Today, most pig owners (especially if they only have a couple of pigs) buy bags of compound food. It is easy to feed and store, and presents less work than mixing your own food. You can also guarantee that your pig is receiving a balanced diet if it is fed in the correct quantities. Compound food comes in different types and for all stages of a pig's life, as follows:

Creep food Comprising tiny pellets, this is usually given ad lib to piglets from three weeks onwards. It is high in protein: at least 18–25 per cent.

Grower food This is larger than creep food, but smaller than sow nuts and still very high in protein. Mix it with the creep food, gradually increasing the ratio daily until you are feeding solely grower food. Start feeding it about two weeks before weaning, so that the piglets are used to it before the stress of weaning. Grower food can be fed right up to the finishing stage, although some finishing units feed specialized finisher pellets.

Sow and weaners' nuts Fed to weaners, these are also ideal for in-pig gilts and for sows feeding piglets.

Ordinary pig nuts Large nuts, usually fed to dry sows and boars, these are suitable for feeding loose on the ground, if it is dry enough.

If the ground is dry and there is no danger of the feed getting lost in the mud, you can scatter feed nuts to encourage foraging.

Meal Powdery flakes that mixed with water form a palatable mash, meal must not be fed dry. It is ideal for sick pigs and those that have lost their appetite, but can encourage pigs to rush their food and even choke. For fussy or sick animals, mix the meal with molasses to make it more attractive.

Feeding amounts

The amount you feed very much depends on your pig and its circumstances. A good starting point is around 0.5–1 kg (1–2 lb) for eight-week-old weaners up to approximately 12 weeks. Thereafter increase this by around 250 g (8 oz) per month until the pigs are ready to go for slaughter.

Dry sows and boars should receive approximately 3–4 kg (6½–8½ lb) a day.

What a lactating sow is fed depends very much on the size of the litter, but you should be thinking of at least 6–8 kg (13–17½ lb) a day – if not more. Watch her condition carefully so that you can adjust the amount accordingly.

Don't be too rigid in your feeding and take the weather into account – if it's cold, pigs should be given extra, as should lactating sows and show pigs. Make sure every pig is receiving its fair share.

Other foods

Do not be tempted to feed pigs kitchen waste, for not only is this illegal in some countries, due to the possibility of spreading viruses and bacteria that might result in diseases such as foot-and-mouth, but you may find that your pigs put on too much weight. However, fruit and vegetables straight from a market or shop or your garden will be much appreciated and should be given as often as possible to supplement your pig's usual food. Don't try to substitute vegetables and fruit for proper pig food, though, or your pig will lose condition and, if it is growing, weight gain will take that much longer. Potatoes must be fed cooked and in small digestible pieces.

Fruit such as apples are acceptable supplements to your pig's usual diet – but do not feed kitchen waste.

Fat levels

Traditional breeds tend to lay down back fat more readily than commercial breeds, so a close eye should be kept on their weight. If they are allowed to get too fat, sows are less likely to conceive. If growers get too fat, you will get back in the meat what you have put in – in other words, the meat will be too fatty. Catch them at an ideal weight, when they have enough flesh but before they lay down too much fat. The chart below will give you some idea of the condition your pigs are in.

Condition	Definition
Emaciated	Bones are clearly visible
Thin	Bones can be felt without pressure when the palm of the hand is laid flat on the skin
Ideal	Bones can be felt only with firm pressure when the palm of the hand is laid flat on the skin
Fat	Bones can be felt only when the fingertips are pressed into the skin
Obese	No bones can be felt

Milk and whey can be given, but in the UK you will need to obtain a licence from your local authority to feed this to your pigs.

As you gain experience with pigs, you may want to mix your own food. There is no doubt that feeding in this way is cheaper, and many top breeders would not feed in any other way. But mixing your own feed must be undertaken with caution. Commercial producers mix a totally balanced feed, but you will be using a lot of guesswork, especially in the early days.

Before embarking on feeding in this way, it is worth researching the different foods available and their nutritional and feeding characteristics, and of course taking advice from other breeders who feed in this manner. Depending on what you are trying to achieve, base foods that are often used in home-mixed feeds include barley meal, flaked maize, wheat meal, fishmeal and soya meal, plus minerals, the most important of which are calcium, sodium, phosphorus and cod-liver oil.

Getting the amount absolutely right, so that your pig looks in the peak of health, takes time. If you get it wrong to begin with, don't worry too much – it's all about gaining experience.

Watering

Pigs should always have access to fresh water and ideally an automatic filling system should be in place for them. The filling bowls should be cleaned daily and checked to ensure they are working correctly.

Watering systems

Galvanized troughs are often used instead of automatic systems because they are easy to clean, reasonably cheap and, being quite shallow, especially suitable for weaners. The disadvantage is that pigs can tip them over quite easily, which could mean that they go all day without water, and in hot weather this could have a detrimental effect on their health. If you use troughs, secure them by wedging them between something solid, such as fence posts. In hot weather, troughs can

Check nipple drinkers daily to ensure they are giving your pigs a sufficient supply of water.

be emptied within a very short space of time, so it is important they are checked at least twice a day and refilled as necessary.

Nipple drinkers are activated by the pigs moving the valve inside the body of the drinker, which releases a small amount of water. The quantity of water flowing through the nipple drinker should be monitored daily to ensure the pigs are consuming adequate supplies. A second supply of water should be offered if there is any doubt about the amount of water that the pigs are able to take.

If you live in a hard-water area, check the pipes regularly to ensure there is no build-up of limescale, which eventually will restrict the flow of water to automatic systems. All pipes connected to automatic systems should be hidden underground or placed behind a secure covering.

Piglets and weaners

Nipple drinkers should not be the sole source of water for piglets and newly weaned pigs. It is far better to offer piglets water in a shallow bowl and to provide weaners with water in a trough.

Piglets should have access to fresh water within a few days of being born, but make sure the dish is shallow enough that they won't drown in it.

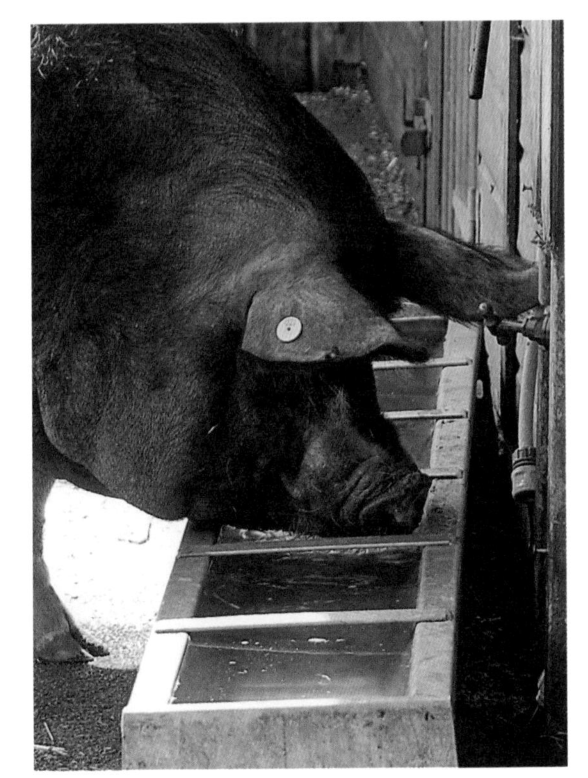

Pigs need a supply of clean water at all times. Any container should be fixed so that the pigs cannot knock it over.

Your pig's daily water requirements

Below are the minimum daily water requirements for pigs, as stated in the UK's Code of Recommendations for the Welfare of Livestock:

Weight of pig (kg/lb)	Daily requirement (litres/pints)
Newly weaned	1–1.5 litres (1¾–2½ pints)
Up to 20 kg (44 lb)	1.5–2 litres (2½–3½ pints)
Finishing pigs, up to 110 kg (242 lb)	5–6 litres (8½–10½ pints)
Sows and gilts: pre-service and in-pig	5–8 litres (8½–14 pints)
Sows and gilts: in lactation	15–30 litres (26–53 pints)
Boars	5–8 litres (8½–14 pints)

Handling

Incorrect handling and a loud voice can cause pigs stress, so they should always be handled in a calm, quiet manner. Handling pigs is the not the easiest thing to do, but the more you do it, the easier it becomes.

Getting to know your pigs

It helps if your pigs know and trust you, and are used to you touching and moving among them. Get into a routine of spending time with your pigs, rubbing them frequently all over their bodies (especially their ears, which they love to have rubbed) and talking to them. Pigs are intelligent and will often come running when they hear your voice, so by talking to them and using their names (if they have them), you are helping to build up a relationship with them that goes a long way towards stress-free handling.

Spend a few minutes each day moving them around in the pen, so that they learn what is required of them and you are safe in the knowledge that they cannot escape. Whenever you do have to move them from familiar situations, do so quietly and try not to make any sudden movements. Pigs that are kept quiet and stress-free will be far easier to move than worried or frightened pigs. When stressed, pigs become excitable, so never lose your temper if they won't go in the direction you want. If you are having real problems, go and get help, or put them back and try again later when you have all calmed down. Bear in mind that pigs with lop ears tend to be quieter but more stubborn to handle than pigs with prick ears such as the Tamworth (see pages 138–141).

Pigs love a good scratch and often go all submissive if you touch the right spot. If they are breeding pigs, get them used to having their stomachs scratched – more often than not, they will roll over and lie down, which helps if you need to encourage the piglets onto the sow.

Moving pigs is much easier if you can teach them to follow a feed bucket, so if it is worth waiting until they are hungry if you need to move them. Pigs will follow buckets of food anywhere, but tread carefully when moving them in this way, as they could move that little bit faster than you and knock you over.

If your pigs are in an enclosure with an electric fence across the gateway, they will probably refuse to come out through the gate – even if you take the electric fence away – so always put a pile of straw across the gateway, as this seems to give them reassurance that they won't get a shock.

Moving piglets

Moving very young piglets is usually not a problem. They will squeal, but they're much more manageable than older piglets and less likely to wriggle. If you do need to move young piglets, carry out the procedure as quickly and quietly as possible, to keep the sow from becoming agitated.

Once piglets are a few weeks old and have started exploring outside their house, moving them can be even more difficult than moving adult pigs. Although they are smaller, so can be picked up and carried, catching them is an art in itself. They are nippier and more slippery than adult pigs, so do not try to catch them in the open – it just won't happen.

If the piglets are in an ark or barn, corner one with a slapboard (see page 24) if necessary, grab it with both hands, placing your hands on either side of its body, and then lift it. If you're not confident about catching piglets, put some food down and wait until their heads

Using a slapboard and pig stick, you can train your pigs from an early age to walk beside you.

are in the trough before attempting to lift them. However, this only works for the first two or three piglets, after which they will be on their guard and will eat with one eye on you.

Try not to be put off by piglets' squealing, which can be quite shocking when heard for the first time. And don't pick them up by grabbing one leg – this risks straining the hocks or pulling joints out of place. Depending on the size of the piglets, some people put them in a large bucket or basket to carry them. This usually has the desired effect of stopping the squealing, but be careful they don't jump out.

Moving adult pigs

The secret of moving pigs successfully lies in the preparation. Just because your pig is a quiet, well-behaved sow, that doesn't mean you can just go into the pen, get her out and expect her to walk across two open fields without any problems. Always expect the unexpected, no matter how experienced you are or how many times you have moved pigs before.

If you are moving your pig from one enclosure to another near by, you need to have at the very least a slapboard, a stick and, for emergencies, that bucket of food. Ideally a temporary fenced walkway using portable gates should be constructed for a trouble-free walk between pens, but make sure it cannot be pushed down by a pig intent on escaping. Check that any gates leading to a road are firmly shut, so that if a pig does escape, it is contained within your property. It is possible to move a few pigs a fair way without using a trailer, but it is certainly not a one-person job. If you are walking them, get enough help to minimize potential problems – far better to have too much help than too little. And don't hurry the pigs along. In summer try to move pigs early in the morning or in the evening when it is cooler.

It is worth training a pig from an early age to move away from a slapboard. Many show pigs seem to move forward and in the required direction without any help from the handler, because they have been taught at a young age what to do when they see the slapboard at the side of their head, and a glancing tap from the stick is often all that is needed to move them forward. The stick should never be used with force, merely as a gentle aid to forward movement. It is worth finding an expert to show you how to use the slapboard and stick correctly.

Trailer-training

All pigs should be trained to go into the trailer without fuss. Store pigs (pigs for fattening) in particular should be able to walk into a trailer quite happily and unafraid. Taking them to the abattoir is stressful enough for the owner, and trying to load an uncooperative pig or pigs on the morning not only adds to your own stress, but also doesn't help the animals.

So, as soon as possible, start trailer-training your weaners. Place the trailer in the enclosure with the ramp down. Make sure the trailer will not tip once the pigs are inside, either by attaching it to a vehicle or by using an axle stand and jack to prevent it from tipping. Ideally, only the ramp should be inside the pen, with the fencing right up the trailer. Put a good amount of straw on the floor and a trough for their food at the back of the trailer.

Generally, pigs only move forward if they are comfortable doing so – gentle perseverance is the key.

Feed your pigs initially on the ground next to the trailer, then gradually move the food up the ramp until they will quite happily go in and eat calmly from the trough. After a while they may even prefer sleeping in the trailer to their ark. If they seem reluctant to go into the trailer, temporarily board up the ark or remove it so that they use the trailer.

If for some reason you don't have a chance to train your pigs early on to go up the ramp for the abattoir, don't panic – getting pigs used to walking up a trailer normally takes no longer than a week, and sometimes they are happy to walk in and out after only a couple of days.

Troubleshooting

Problems in getting your pigs into the trailer can still arise and, if you are trying to load an adult pig that refuses to budge or keeps backing off, the situation can become quite fraught. If you do not have a walkway that can be closed off at one end, with the trailer backed into the other end, you need to look at other ways to minimize potential problems.

One method is to park the trailer as close to the pen as possible and up against something solid, such as a wall; this means that, once the ramp is down, you only need to worry about making one side and the back escape-proof. You will need portable gates to fashion a short walkway up to the trailer; if possible, these should have some sort of covering attached to them, such as plywood or cardboard sheets, to prevent the pigs from seeing through them. Secure the portable gates to the trailer side-gates with bailer twine. Find at least one other person to help with the loading and give him or her slapboards.

Pigs are very strong, so encouragement with food should be tried first. Encourage your recalcitrant pig into the walkway with a bucket of food and, once there, get your helpers to stand behind it with their slapboards to prevent it from turning or backing. Stand in front of it, encouraging it forward with your voice and with food, while the helpers at the back persuade the pig to move forward. Take your time, for you want to keep the pig as calm as possible. Not moving at all – although it is infuriating – is better than moving backwards at great speed.

Generally pigs only move forward if they are comfortable doing so. It is therefore imperative that the ramp is stable and won't rock once the pig has stepped on it. If it moves, your pig will be startled and will back off the ramp, making it harder to reload. You may find that putting straw in the walkway and on the ramp helps to persuade the pig to move forward into the trailer, as does laying a trail down the ramp of food that it likes, such as carrots or apples. At all times encourage it with your voice and never get angry. Loading one pig (or even two) can often be harder than loading several. Store pigs tend to follow one another, once you have one going up the ramp.

Safety warning

Never position yourself in front of or at the side of a boar's face. Boars have tusks and can cause considerable damage just by tossing their heads. If you have to stand close to a boar, make sure you are behind his shoulders.

Handling boars

By nature boars are more aggressive than females, so they should always be handled with respect and firmness, no matter how quiet they seem. Never trust a boar completely, no matter how well you know him. When you are moving him around, always keep one eye on what he is doing and where he is. Be aware that boars are a lot bigger and more substantial than sows, so one could hurt you even without meaning to by knocking you to the ground. If you have sows near by, especially if they are in season, a boar's personality could change without warning, so it is always a good idea to have a slapboard with you or at least within easy reach.

Establish the ground rules early on concerning what is acceptable and what isn't. It may seem like a bit of fun for a young pig to nibble your boots at a few months old, but what happens when he's a fully grown boar and takes a bite instead of a nibble? Don't allow a boar to barge into you, especially at feeding time. One day he could push you over, and you are then in a vulnerable position. However, don't be frightened of your boar: give him the same amount of attention you give the other pigs, to build up a good relationship with him. Just like the others, he will enjoy being talked to and given a good scratch and a rub.

When moving your boar from pen to pen, it is worth always having someone to help you, especially if you need to take him past other pigs. Walking past other boars is to be avoided at all costs.

Hygiene and bio-security

To help prevent the spread of disease and infection among your pigs, strict hygiene and bio-security measures should be taken on a day-to-day basis. Some simple guidelines are given below.

Basic measures

Although pigs kept outside are hardy and not usually prone to illness, infection can spread among them quite quickly. A few simple bio-security measures should be in place to keep serious diseases at bay and, on a day-to-day level, shelters, feed and water utensils should be kept as clean as possible.

Ideally you should have a small washroom (or at the very least a sink with hot water) close to the pig pens. Failing that, use an anti-bacterial handwash as a matter of course after handling or feeding the pigs.

Safe practices should start at the farm gate. It is worth investing in a mat impregnated with specialist disinfectant. Place it at the entrance to your property so that vehicles have no choice but to drive over it. Choose a mat large enough for all four wheels to have at least one full turn on. Bear in mind that this will only disinfect the wheels, not underneath the car, and that you will have to top it up with disinfectant regularly.

As part of your bio-security it is a good idea to position a footbath for visitors by the pig pens. Keep the disinfectant topped up and, if it rains, try to cover the footbath, as water will weaken the solution. The footbath should be deep enough to completely submerge the feet of rubber boots. A stiff brush should also be on hand to scrub the soles of boots, especially if they are ridged.

Pen hygiene

On the whole pigs are clean animals, but every now and again they will dirty their ark or shelter rather than go outside, especially if it's very cold or wet. Once they get into the habit, it is very difficult to break them out of it and you may have to consider moving them to a different pen to stop them. If your pigs are kept in a barn or large shelter other than an ark, make sure the space inside is big enough to have a clearly defined area for both sleeping and dunging, though sometimes (especially with weaners) having too much space will actually encourage them to soil wherever they feel like it. If you think this is the likely cause, try restricting them to a smaller sleeping area, with access to an outside area for dunging.

Bedding should be kept as fresh as possible and free from contamination by rodents. Don't use the bedding if it has rodent droppings in it or is musty and stale – it is better to burn it and use other bedding instead. Bedding should also be changed regularly, especially in winter when the likelihood is that the pigs will drag in mud. In summer – unless there is a reason to clean it out more often – bedding should be taken out and burned and the shelter disinfected once a month. If your pigs are kept in arks, a removable floor enables you to take the top off completely and give the floor a thorough clean. Plastic arks can simply be rolled on their side and pressure-hosed. Be very careful when cleaning arks with fixed floors not to leave any soiled bedding in the corners.

It is essential that farrowing sheds are thoroughly cleaned out and disinfected every week, and certainly between the existing occupants moving out and new occupants moving in. If possible, power-wash all the farrowing areas. All evidence of the afterbirth must be removed as soon as possible and burned. If possible,

Arks with detachable floors are far easier to clean than the fixed-floor arks.

You will need to thoroughly disinfect your ark or other housing at least once a month.

farrowing arks should be moved to a new area after cleaning. If you are handling more than one litter it is worth using disposable gloves, otherwise wash your hands thoroughly between litters.

Disinfectants

There are so many different farm-related disinfectants on the market today that it can be confusing to know which one to use. Ask suppliers for advice as to which one is best for your intended purpose. Smallholders usually keep at least two different disinfectants: a general one for day-to-day cleaning, and another one for the footbath. It is also handy to have a powder disinfectant, which can be used on bedding in the sleeping quarters. Whichever disinfectant you buy, make sure that it is harmless to animals. If it is not, then the treated area must be left for sufficient time (in accordance with the manufacturer's instructions) before the pigs are allowed near it.

Feed and water utensils

Utensils used for feed and water must be kept clean. Food that has not been eaten within an hour must be taken out and thrown away. If left, it will encourage rodents and birds around the pens. Birds can spread Erysipelas (see page 72) and other diseases through their droppings. Rinse utensils daily to remove bits of feed stuck to the side. Two or three times a week give them a good scrub with hot soapy water, getting right into the corners, then rinse them thoroughly before putting them back into the pens. Particular attention should be paid to automatic drinkers, which should be free from debris and mud to ensure the free flow of fresh water at all times.

Wherever you keep feed, you will attract mice or birds, so it must be stored in containers in a rodent-proof building. Ideally windows should be kept open to allow the free flow of air, but should have a mesh-screen to prevent birds from entering. Any containers that are holding loose food should be dry and clean.

Isolation pens

An isolation pen (of a type approved by the government department concerned with animal welfare) is useful if you have the space. It enables you to bring animals onto your property without incurring a 21-day standstill (see page 45), and is ideal for those owners who show throughout the summer or are continually moving pigs on and off their property. More importantly, it means that any new pig brought onto your farm can be isolated and watched for signs of illness, without risking the health of your other pigs. Overalls and rubber boots should be changed after leaving the isolation area.

A pig that falls ill should always be isolated. If you do not have a separate isolation area, at least try to move it out of its pen and away from the other pigs – a stable or barn will suffice to stop infection spreading. Wash your hands thoroughly after touching the sick pig. Place a footbath (see page 62) at the entrance to wherever the sick pig is being kept and walk through it before and after handling the pig. All bedding should be burned after use and the shelter thoroughly disinfected. Any troughs or buckets that have been used should also be washed out with hot soapy water. Do not return the pig to the others until the infection has passed and your vet confirms that it is safe to do so.

Visitors

Visitors to your property pose the biggest risk of spreading infection. Equally, you have a responsibility towards your visitors to ensure that they do not pick up any infection during the visit. As they will not be familiar with your hygiene routines, they should be accompanied at all times around the pig pens. It is worth putting up a 'No entry' sign to discourage unsolicited and unaccompanied visits.

If the visitors are coming to inspect your pigs, make them aware before arrival that their clothing and footwear must not have been in contact with other farm animals within the 48 hours prior to their visit. Ask them to park their car as far away from the pig pens as they can. If they have a dog with them, it must stay in the vehicle or be taken for a walk well away from the pens. Make sure that their rubber boots are free from mud before they step through the footbath, because footbaths only work if boots are clean. On leaving, visitors should be given access to a sink in which to wash their hands or to a antibacterial handwash.

Pests and Diseases

Pigs are healthy creatures and rarely ill, but if they do pick up an infection, they can be pulled down quite quickly. It is helpful to know how to spot common problems, how to deal with them yourself and when to summon expert help.

The healthy pig

Traditional pig breeds are hardy and usually enjoy good health. To keep them in peak condition, you'll need to give them plenty of exercise, fresh air, adequate food and a sturdy shelter.

The healthy pig

The owner can look for various indicators in a pig's physical appearance and manner to check that the animal is healthy.

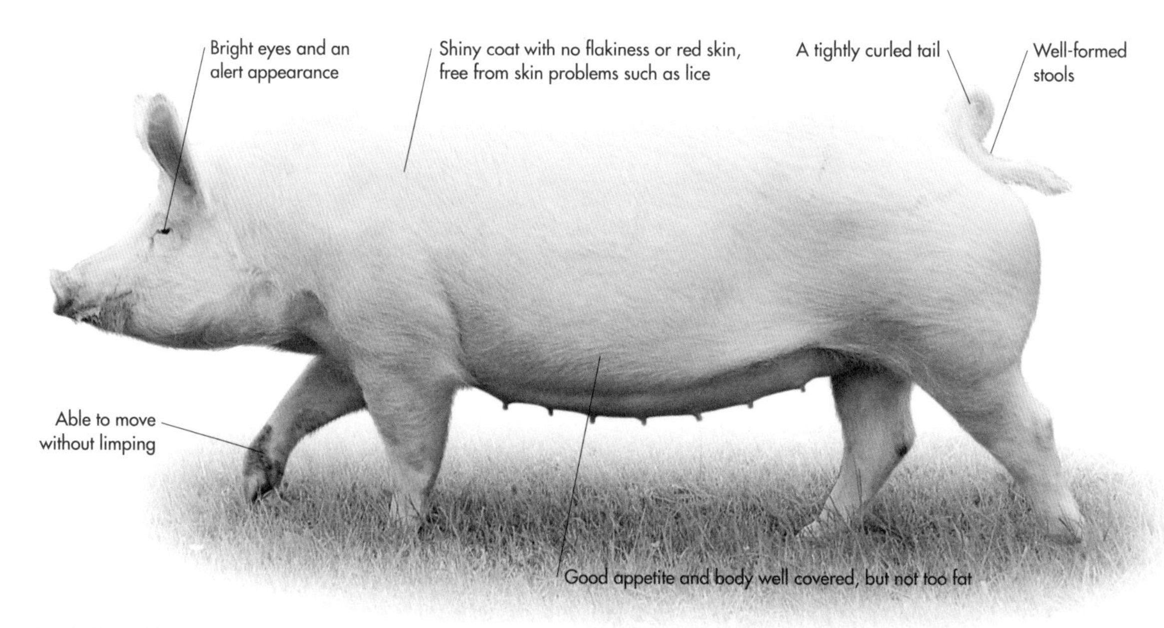

Basic health matters

Most pig owners know when something isn't quite right. Occasionally pigs will go off their food or go lame for no obvious reason, and just as quickly they will recover. However, if symptoms (even mild ones) go on for longer than a couple of days, summon your vet.

Loss of appetite can occur if the weather is hot or the pig is in season. Rather than continuing to put your pig's usual food out every day, try tempting it with other food, such as fruit or vegetables. Pigs love bananas, and many a sick pig will prefer them to other food. Make sure you do not give your pig any illegal foods, such as kitchen waste or meat products. Often, pouring milk over the usual feed will tempt the pig into eating again, although you may have to register with your local authority as a milk user.

You take the temperature of a pig by inserting a thermometer in the rectum. A healthy pig's temperature should be 38.6–38.8°C (101.5–102°F). Higher than this usually means that the pig has a fever, which can be lowered by a course of antibiotics. Lower than this, and the pig should be kept as warm as possible using a heat lamp and blankets.

Pigs should be wormed regularly (see page 71). Depending on your farming practices, you may wish to vaccinate, although if you are organic neither worming nor vaccination may be permissible.

Injections

To save on vet's fees, it is worth learning how to inject. The skin of a pig can be quite difficult to penetrate, so you need to approach injecting with a positive attitude.

There are two ways of injecting a pig: into the muscle (intramuscular) or under the skin (subcutaneous). Muscle injections are easier and quickest to do. When injecting, ensure that you use the correct gauge needle. In general, a 16-gauge needle should be used on adult sows and boars; a 21-gauge needle on piglets up to about 10 kg (22 lb); and a 16–19-gauge needle on older pigs of about 60 kg (130 lb). Needle size also depends on the type of medicine, so consult your vet.

Inject approximately 5–7.5 cm (2–3 inches) behind the ear, at the point where the loose skin borders the taut skin. If the injection is too far back, you might inject into fat, which will lower the pig's immunity response. For subcutaneous injections, pinch a fold of skin just behind the ear and insert the needle at an angle, to ensure that you inject under the skin.

Pigs must be well restrained when injecting them. Try not to stand the pig on straw, because if you do drop the needle, it will be hard to find. If you feel that you may have problems keeping the pig standing still, consider using a pig twitch (see page 28). However, most breeders inject while the pig is eating – more often than not, it will simply grunt and carry on eating.

Ensure that all needles are sterile before use and that, once used, they are disposed of correctly and in accordance with your country's current waste regulations.

Bad signs

It is important that the pig is checked every day so that even minor problems can be dealt with quickly. Some of the signs that all may not be well are as follows:

- Lameness
- Wounds
- Flaky or angry, red skin
- Vomiting
- Coughing
- Rapid breathing
- Standing hunched up
- Loss of appetite or thirst
- Change in stools
- A general air of looking miserable

Withdrawal periods

Most licensed medicines have a withdrawal period (usually 28 days), during which time you are not allowed to use the animal for food production. In many countries, forms have to be filled in when an animal goes to slaughter, confirming that no medicines have been given during this time. Natural products such as aloe vera (see page 76) do not have a withdrawal period, but you should still keep a note of any such products that you have given your pig.

If you suffer the misfortune of a pig dying on you, the carcass must be disposed of in the correct manner and in accordance with your country's regulations. Burying or burning on your property may not be permitted and you may have to arrange for your pig to be picked up by a licensed collector for disposal elsewhere.

Health problems

There are a number of health problems that can affect your pigs. Knowing a little about how to spot them and what remedial measures to take may help prevent them from becoming serious.

Common problems

Common problems range from heatstroke to mange to lameness. Some can be treated by the owner if caught early enough, while others require a vet's intervention.

Heatstroke

Heatstroke is quite common in pigs and occurs when the pig is unable to lose body heat and its temperature rises dangerously. In hot weather provide as much shade as possible, either in the form of trees or by erecting a pig shade. Ideally, arks should be insulated and ventilated to allow air to flow through. If it is very hot, replace the straw with something less insulating or take it away altogether. Ensure that buildings used for farrowing sows are kept cool and that the air flow is not restricted. If you are travelling with pigs, make sure all the trailer vents are down; dampen bedding in the trailer if the weather is extremely hot.

All outdoor pigs must be provided with a wallow in hot weather. The best ones are man-made wallows containing mud, which retains the moisture on a pig's body far better than water in a galvanized steel wallow.

Once a pig has heatstroke, its temperature must be brought down as quickly as possible, but not in such a way that it causes shock to the animal. Never throw cold water over a hot pig, as this could result in death. Sponging cold water over the head and body, as well as spraying the air surrounding the pig and turning on a fan, will assist in bringing the pig's temperature down. Vinegar dabbed behind the ears can also help to reduce the temperature; always take some with you when you go to a show or when travelling with a pig.

Treat heatstroke by gently sponging the pig with cold water to bring down its temperature.

Signs of heatstroke

- Excessive salivation, sometimes with blood mixed in with the froth
- Rapid breathing and gasping
- Trembling limbs
- Staggering and uncoordinated movement

Sunburn

It is mostly pale-skinned pigs and young pigs that are susceptible to sunburn. On the whole, because of their darker colours, traditional breeds do not suffer from it.

Avoid sunburn through preventative care, similar to the measures used against heatstroke. Wallows should be provided as well as shade (or keep your pigs in during the day and let them out at night); rub a high SP-factor sun cream, of the type worn by humans, onto pigs that may be at risk (not forgetting snouts and ears). If your pigs do get a touch of sunburn, dabbing on calamine lotion will ease the discomfort.

Worms

Pigs should be wormed routinely – usually at eight weeks, and then about every six months afterwards if you are keeping them for breeding. In-pig sows and gilts should be dosed a couple of weeks prior to farrowing, and again when the litter is weaned. All new pigs that arrive on your property should be wormed before being allowed to join the other pigs.

Worming must go hand-in-hand with good paddock management. Pigs become infected when eating or drinking, while piglets usually pick up worms by suckling. Enclosures that have had pigs on them for a long time will carry a heavy burden of worms, so if possible it is good practice to regularly rest and crop-rotate the paddock.

Slaughter houses will reject pig livers that have been damaged by worms, which is known as 'liver spot'.

Worming can be controlled by injecting the pig with an anti-parasite product or by means of an anthelmintic (parasite-expelling) preparation given in the food. Injections tend to be more reliable because you can be certain the pig has received the correct dose.

E. coli

The *E. coli* bacterium is usually the reason for high mortality rates in piglets, but infection can be prevented. Veterinary assistance should be sought

Providing a wallow for your pigs in summer will help to keep them cool and avoid heatstroke or sunburn.

Symptoms of worms

- Failure to put on weight in growers
- Scours and coughing in young piglets
- Loss of condition
- Staring coat (dull and lifeless)
- Pot bellies

quickly once *E. coli* is suspected. Sows can be vaccinated on a regular basis and boosters given prior to farrowing; this immunity is then passed on to the piglets via the mother. Although *E. coli* is a major problem in large pig units, even the small-scale pig-keeper can be affected by it, so vaccination is well worth considering. Good hygiene – especially in the farrowing area – also goes a long way towards keeping *E. coli* at bay.

Scours

Scouring (diarrhoea) can occur in adults and young pigs. Sometimes it is temporary, brought about by a change of food, worms, overeating or too much fruit or vegetables. However, if it goes on for too long, it can cause dehydration and even death in piglets. Scouring in piglets often occurs at weaning so they should be carefully watched to ensure that it is only temporary and not a bacterial infection, which can kill a piglet very quickly. Piglets that have been hand-reared often get the scours, as do piglets that have not received enough colostrum (first milk feed, see page 90).

For scouring pigs, a proprietary powder that is soluble in water is an effective way of drying them up. Any shelters that have been used by the affected pigs should be cleaned and disinfected and the bedding burned.

Erysipelas

Erysipelas is not as common as it once was and, thanks to vaccination, good hygiene and better awareness, it can easily be prevented. The bacterium that causes erysipelas can be carried by birds and rodents and is found in the soil. It is possible (but rare) for humans to catch it through cuts, so gloves should always be worn if erysipelas is suspected. The organism can survive for many months if the conditions are ideal, and is far more common in straw-bedded systems than in fully slatted units.

Erysipelas can cause reproductive failure in sows and lameness in younger pigs. Abrasions appear on the skin, often in the shape of diamonds, and pigs experience a high temperature and look extremely miserable. Other symptoms include lameness or stiff gait and an arched back.

If caught early on and treated with antibiotics, the rate of recovery is usually 100 per cent. However, death is likely to occur if the early signs are not picked up. It is therefore vital to vaccinate against this disease. Sows are most susceptible to it in the later stages of pregnancy. They should be vaccinated and then given a booster three weeks before farrowing, which will pass some immunity to the piglets. Boosters should be given to all pigs (including boars) every six months.

Meningitis

Meningitis can be brought on by a bacterial infection or by stress, and pigs that suffer from it rarely recover. The first signs are depression and a reluctance to stand. As the disease progresses, convulsions take place and once this happens death will occur within a few hours.

For the animal to have any chance of recovering, antibiotics must be given rapidly. The pig must be removed from its pen and put in an area with deep bedding to prevent injury. Fluids containing electrolytes to counteract dehydration should be given, although they may have to be tubed if the pig is reluctant to drink. If meningitis is caught in time, recovery is swift. However, if no change has occurred within 48 hours, the pig will probably have to be put down.

Lice

Lice on pigs can easily be seen with the naked eye. They look like tiny crabs and can cause great discomfort to the pig, often grouping around the ears before spreading over the rest of the body. The louse has a lifecycle of around three to five weeks on the pig's body; off the body, it can only survive for a few days. It lays its yellow eggs close to the skin, normally around the ears, jowl, belly, flanks and shoulders. Symptoms that your pig has lice are continual

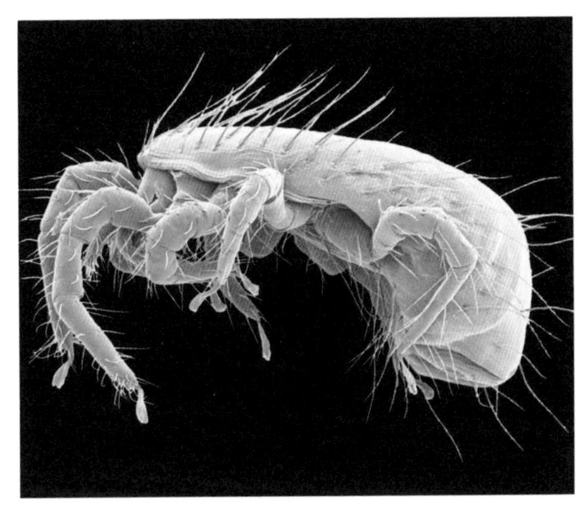
The sarcoptic mite is the most common cause of mange, burrowing under the pig's skin to lay its eggs.

Looking like miniature crabs, lice live on a pig's skin and can be spotted with the naked eye.

scratching and dry, flaky skin. Sucking lice can result in anaemia if treatment is not given in time.

Lice can be controlled by injecting pigs with an anti-parasitic solution and using an anti-parasitic wash. If your herd is free from lice, pigs that are brought onto the property should be checked thoroughly before they are allowed near the others.

Mange

Mange is a skin disease caused by the *Sarcoptes scabiei* or *Demodex phylloides* mites and is usually brought in by a carrier pig. By far the most common is the sarcoptic mite, which burrows under the skin to lay its eggs, making the pig extremely uncomfortable. Mites favour the head and ears, and signs of mange are head shaking, redness and a crusty, sometimes bloody skin, plus continual scratching and rubbing. If the mange is allowed to continue, it will lead to loss of condition and possibly death.

The mite is capable of living off the pig for two to four weeks if the conditions are ideal, so this should be borne in mind when trying to eradicate mange in a herd. Injections of anti-parasitic solution have proved extremely successful, while some breeders use a water solution of tea-tree oil (see page 76) with good results.

Colds and pneumonia

Just like humans, pigs can catch a cold; the first sign is usually a runny nose and loss of appetite. The pig should be kept warm and in a ventilated shelter. If it will eat, feed it as normal. If it has lost its appetite, tempt it with warm mashes and succulent greens.

Pneumonia can pull a pig down rapidly and, once suspected, veterinary assistance should be sought if death is to be avoided. Signs of pneumonia (as opposed to a cold) are coughing and rapid breathing, plus a high temperature, diarrhoea and loss of appetite. Growers that have had pneumonia in the past and recovered often fail to put on weight as quickly.

The spread of the disease is either airborne or by pig-to-pig contact, so the affected pig must be isolated from other pigs until it recovers. Pigs can often be carriers of viral pneumonia and not show any signs. Antibiotics will be required to treat the affected pig.

Joint ill

Joint ill is caused by bacterial infection entering the body of the piglet, usually through the navel when it is born. However, it can also enter the body through abrasions on the skin, such as knee wounds caused by suckling, or occasionally through clipped teeth. Once in the body, it spreads to the joints, causing lameness and stiffness. The first sign that something is wrong may be that a piglet is reluctant to stand or appears lame; or swelling might appear in the affected joint and the piglet might squeal in pain when it is picked up.

For the piglet to recover, it is important that antibiotics are given as soon as possible. Painkillers may also be needed as joint ill can be very painful.

To prevent joint ill occurring, piglets' navels should be dipped or sprayed with iodine, and all farrowing areas should be kept clean and disinfected. Bedding should be deep enough to prevent abrasions on the knee, and high standards of hygiene should be in place.

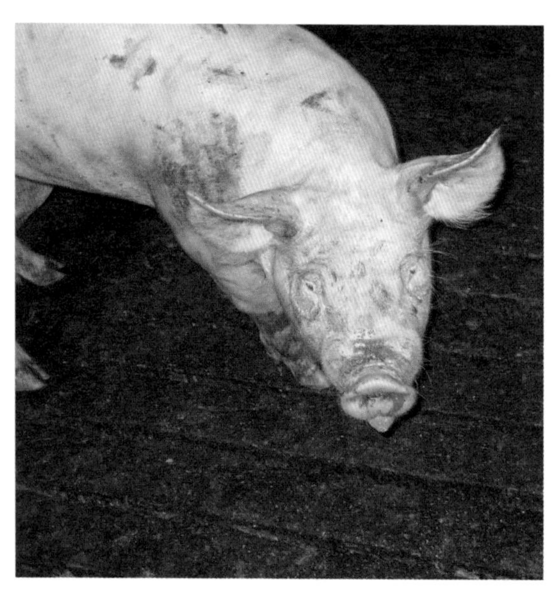

An inflamed twisted snout is one of the symptoms of a severe case of atrophic rhinitis.

Gastric ulcers

It is thought that stress plays a part in pigs being affected by ulcers, which often occur in lactating sows. Ulcers are difficult to diagnose. The signs vary, depending on whether or not the ulcer is bleeding. Occasionally there is vomiting and the pig will appear tucked up (with a tight stomach). Death often occurs and sometimes this can happen suddenly and without much warning; pigs can recover, but this is rare.

Iron deficiency

If piglets don't receive sufficient iron from the sow, they can suffer from iron deficiency and become anaemic, although piglets with access to soil do not normally suffer from this complaint. Injecting piglets with iron within a few hours of being born is normally the course of action recommended by vets, but this can bring on the scours (see page 72). Most breeders put sods of earth in the creep area, which is just as effective.

Atrophic rhinitis

Rhinitis is a respiratory disease that results in the tissues of the nose becoming inflamed. It is transmitted either through the air or pig to pig. This disease can be found in two forms: the more common mild, non-progressive form, and the far more serious disease, progressive atrophic rhinitis (PAR). With the lesser disease, infection and inflammation usually occur over a period of around three weeks, but do no lasting damage.

Symptoms are usually apparent at three to eight weeks of age. Initial symptoms are sneezing, sniffling and a nasal discharge. If the disease progresses to PAR, the snout becomes distorted and wrinkled and starts to twist sideways or upwards. Pigs that have only been affected mildly by the disease will not show this distorting of the nose, but may be more prone to bronchitis and pneumonia.

It is possible to vaccinate sows and gilts against rhinitis, if the disease in present in the herd, but this is not recognized as a wholly effective preventative measure.

Poisoning

Pigs will eat anything, so take this into consideration when putting them in an enclosure for the first time, although most plants that are poisonous to them have an unpalatable taste and are usually left alone. Some plants will bring death in a matter of hours when ingested; others will simply make the pig very ill, but it will recover. Some common causes of poisoning are given below.

- **Bracken (*Pteridium aquilinum*)** Pigs have been kept in woods with bracken for centuries; indeed, they are often used to destroy it. But bracken fronds contain a toxin that induces a vitamin B deficiency and can prove fatal. However, it is generally thought that as long as the pigs have access to fresh water and are fed their regular pig food, they should suffer no ill effects from being among bracken.
- **Acorns** The high tannin content in acorn shells (the fruit of the *Quercus* family) can cause abdominal pain, diarrhoea and abortion in sows. Pigs should not be kept in areas where there are large quantities of acorns, especially those pigs that are not used to them.
- **Ragwort (*Senecio* var.)** This is a highly poisonous plant, which should be pulled up, removed from the enclosure and burned.
- **Rhododendron (*Rhododendron* spp.)** Both the leaves and flowers of this plant are toxic to pigs; the roots are non-toxic.
- **Laburnum (*Laburnum x watereri*)** Pigs that have eaten laburnum will suffer convulsions and death will soon follow.
- **Foxglove (*Digitalis purpurea*)** This has an unpalatable taste and is therefore rarely eaten by pigs. If it is ingested, the pig will slip into a coma and death will follow.
- **Deadly nightshade (*Atropina belladonna*)** Initial signs of poisoning are trembling followed by convulsions and then death.
- **Elder (*Sambucus nigra*)** Although elder poisoning is rare, there have been reports that pigs have died from eating the leaves.

Swine pox

This is caused by the swine-pox virus and is often linked to external parasites such as lice. It is highly infectious, and any pig that has swine pox should be isolated straight away, its bedding burned and its shelter thoroughly disinfected. Gloves should be worn when handling the pig and another set of clothing worn if having close contact with other pigs in your herd.

Initial signs of swine pox are areas of redness that quickly develop into a scab, which soon blackens. Swine pox may affect the whole body and there is no treatment, although the pig recovers quite well as long as there is no secondary infection of the scabs.

Mastitis

Mastitis is a painful inflammation of the udders that occurs just before or after farrowing. It has a number of causes, such as an *E. coli* infection, lying on cold floors or injury to the udders by the suckling piglets.

Mastitis causes the udder to feel hot and hard, and it will be painful to the touch. Often the sow will go off her food and have a high temperature. If mastitis is severe, the sow will not let piglets suckle. Treatment is usually antibiotics. A good standard of hygiene and management is necessary to prevent mastitis.

Agalactia

Sometimes sows do not let down milk and there is no obvious reason for this, but an injection of the hormone oxytocin should get the milk flowing. Agalactia, or lack of milk, can sometimes be due to an incorrect diet.

Lameness

Surprisingly common in pigs, lameness can be brought on by a variety of causes, including injury, arthritis, inappropriate flooring or bacterial infection. Some it can be prevented, especially when it is caused by lack of bedding. Occasionally it is caused by poor skeletal structure, so when purchasing a pig you should inspect its conformation and how it stands and walks.

Flooring (especially concrete) can increase the risk of lameness, so ensure that all floors (including the trailer floor) are not slippery when wet. Lameness sometimes occurs through other types of injury, such as fighting or being trodden on, so prevent this as far as possible.

If there is no apparent cause for the lameness, a bacterial infection should be suspected and the vet should be summoned.

Notifiable diseases

If you suspect your pigs have a notifiable disease, it must be reported to the veterinary manager of the relevant government department (such as Defra in the UK) or your nearest animal-health office immediately. All precautions should be taken to isolate the animals and prevent visitors from entering.

Foot-and-mouth

Foot-and-mouth is a highly contagious and notifiable disease and, if suspected, it should be reported straight away. It is caused by a virus, and symptoms include lameness, salivating at the mouth, a very high

Natural medicines

In many countries, natural remedies are used as a matter of course if conventional ones are not available. Extracts of tea-tree oil, tansy, lavender and eucalyptus have had considerable success in the treatment of skin diseases and wounds. Vets use licensed medicines automatically and often have only a limited knowledge of alternative medicines. If you are interested in using natural medicines, you will therefore need to approach specialist companies for advice on the suitability of their products. Two of the more popular natural remedies are:

- **Tea-tree oil** This comes from the leaves of *Melaleuca alternifolia*, a plant native to Australia. It has long been used as a wound-healer by the Aborigines, and its value as an antiseptic is recognized worldwide. The two active chemicals in tea tree, terpenes and cineole, are well known for their ability to destroy most bacteria, fungi and yeast, and the oil is of great value in treating skin diseases and parasitic infestations.
- **Aloe vera** The *Aloe vera* plant contains more than 200 naturally occurring nutritional substances and possesses a number of health benefits. Aloe-vera gel applied locally encourages skin regeneration and can be used directly on wounds, sunburn and insect bites.

temperature and blisters, possibly on the snout, tongue and heels. Pigs can recover from it, but if foot-and-mouth is confirmed the animal will be slaughtered.

Classical swine fever

Good bio-security measures are required to prevent this highly contagious disease. The virus survives for a long time off the pig and can be transmitted by vehicles, equipment and footwear, as well as by infected pigs and pig meat. Symptoms are constipation followed by diarrhoea, coughing, blotchy discolouration of the skin and weakness of the pig's hindquarters. In its acute form, there is a high mortality rate.

Exotic diseases

The serious diseases African swine fever and anthrax are less frequently encountered in Western herds.

African swine fever

This highly contagious disease is caused by flies and ticks, as well as by direct contact with infected pigs. The incubation period for the virus is between five and 15 days, and the first signs are a lack of appetite and a high temperature, followed by coughing, diarrhoea and a darkening of the skin, especially around the ears and snout. The pig will appear weak and reluctant to stand.

Anthrax

Caused by a bacterial infection, anthrax is often found in South America, Asia, Africa and the Middle East. Anthrax in pigs is characterized by nausea, appetite loss, swelling around the neck, severe diarrhoea and often vomiting of blood. The disease progresses rapidly and treatment by antibiotics must be swift to prevent death, although this is often how it ends. It is possible to vaccinate against this disease in some countries.

When to call the vet

Over time you will get to know your pig and recognize what gives cause for concern. It is time to call the vet

Experience will teach you when your pigs are behaving out of character and it is time to call in the vet.

when your pigs are not behaving in their usual manner. You may be used to them going off their food when they are in season, but if lack of appetite is coupled with your pig looking tucked up and miserable, there is something wrong and you should summon help.

Some skin problems, such as lice (see page 72) and mange (see page 73), can be dealt with without the intervention of a vet. Wounds that are not deep or gaping can be eased by a spray of antiseptic. Every pig owner should know how to inject, even if just for worming purposes and if antibiotics are required; calling the vet out every time you need to inject your pigs will be a costly business.

Spend some time researching the various illnesses and symptoms that affect pigs. If you have a general understanding of what diseases and ailments a pig can suffer from, you will be in a better position to know when to summon expert help.

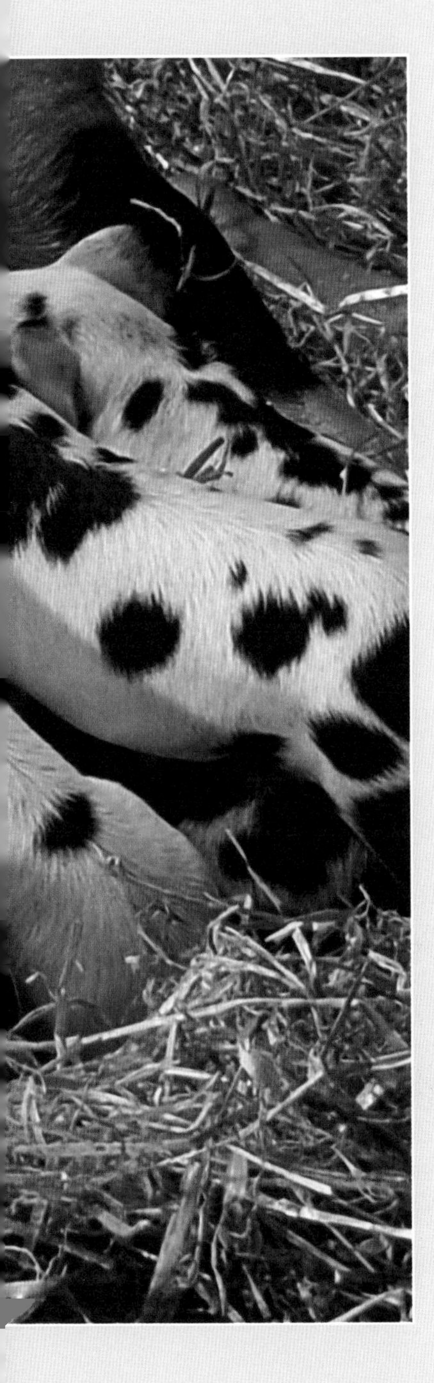

Breeding

No matter how many times you have bred pigs, you cannot fail to be moved by seeing piglets born. This chapter explains how to select your boar or sow and prepare for the birth. Watching the piglets take their first hesitant steps and hearing the contented grunts of the sow will be the reward for your efforts.

Selecting a boar or sow

Foundation stock (breeding stock) should be chosen carefully, otherwise you will be seeing the faults in future litters for years to come. Whether you are buying to breed for meat or for the show ring, invest in the best boar or sow you can afford.

Researching bloodlines

It is worth spending a little time beforehand researching the bloodlines of your preferred breed and, if possible, visiting pigs of this line. Some bloodlines are known for certain faults; other bloodlines are rare and difficult to get hold of. However, be cautious when buying rare bloodlines – they could be rare for a reason.

Always buy pigs that are registered as pedigree. They should come to you earmarked in one of the three acceptable ways, by tagging, notching or tattooing. If a pig isn't earmarked in one of these ways, the chances are it is not registered. Pigs that are deemed suitable as breeding stock should be either herd-registered or eligible for herd registration.

Weaner or show pig?

When buying breeding stock there are two options open to you and what you buy depends on your budget. Weaners are the cheapest option, although there are drawbacks to buying pigs so young. You do not know how the weaner will turn out and having parents as champions is no guarantee that the offspring

Temperament is an essential factor to take into account when buying a boar.

will follow in their footsteps – although good genetics tend to be passed on, this might not always happen. You also cannot guarantee a weaner's temperament as it grows up. Lastly, you will have to wait until it is at least nine months old before putting it to a boar. Boars can work earlier, but you will still have to wait until he is at least seven months old and then you will need a gilt of a similar size.

If your budget is higher, you could buy a show pig. A good place to start looking is at agricultural shows that hold pig classes. Breeders nearly always bring pigs to the show that they have bred themselves, and although the stock they bring will clearly be the best of the litter, it will give you a good indication of the type of pig they are breeding. If they have brought more than one pig (as is often the case), you will get an even better idea of the overall quality their breeding stock is producing. Often there will be an information board detailing litters for sale from the same parents as the show pig, and occasionally the breeder makes it known that the actual pig is for sale after the show, so it is worth checking to see whether this is a possibility if an animal catches your eye and ticks all the right boxes. However, as a show pig will be that much older than a weaner, expect to pay top dollar for it. On the positive side, you may not have to wait long before it can be mated; if it's a gilt, the breeder will sometimes even put it in pig for you, although obviously that will affect the cost.

What to look for

Faults are often hereditary (especially underlines – the lines of teats underneath), so it is important that both boars and gilts have the required number of teats, which should be evenly spaced and of a similar size; and on a gilt the teats should start reasonably well forward. Whatever faults a sow or boar has with its underline, it is likely these will be passed on to the offspring, so be absolutely certain that the underline looks as it should.

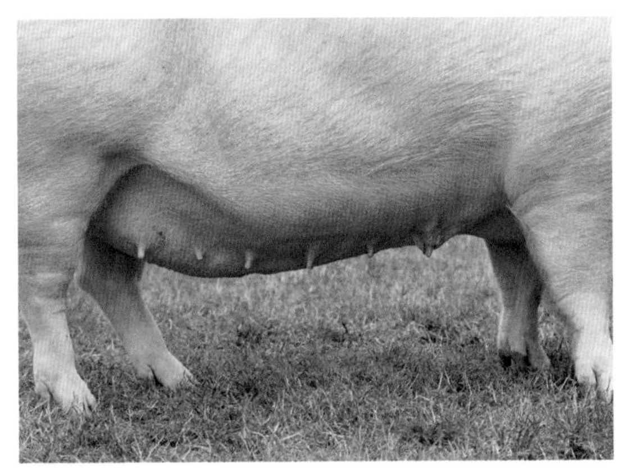

When choosing breeding stock, it is essential to look for a good underline – evenly spaced teats of a similar size.

Avoid buying the runt of the litter or pigs that have crooked jaws or legs. And although mismarked pigs do not strictly have a conformation fault, they should also be avoided, due to the possibility of passing the fault on to the offspring. Saddlebacks (see pages 128–131), for instance, should have a nice clean, even saddle and will be marked down in the show ring if the saddle is anything less. If you plan to breed for meat, buying mismarked pigs is not as problematic as if you are hoping to sell the offspring as breeding stock. However, plans can change over time, so it is usually better to buy a quality animal with the correct conformation and markings in the first place – that way you will, hopefully, have the option to sell your weaners as breeding stock or meat pigs.

It is vital that both boar and gilt have a good temperament – especially boars. They are incredibly strong and can be unpredictable at the best of times, so make sure their temperament is good and find out about the temperament of the parents if you can.

If you are inexperienced, try to take along an expert who can advise you and who may notice things that you will miss. It is worth taking trouble to avoid faults, as the chances are these will reappear in the litters.

Farrowing equipment

Farrowing sows rarely suffer complications, so there is no need for lots of equipment. However, breeding is not the sort of project that should be entered into lightly and certainly not unless you have the right facilities.

Planning ahead

Each sow you breed from will give you a litter that requires your attention from birth until at least eight weeks, and possibly longer if they haven't sold or you are planning to raise them for meat. So make sure at the outset that you have sufficient land and housing to accommodate them all.

Imagine the worst-case scenario: your sow could have at least eight piglets in her litter and, if you haven't found homes for them when they are weaned, you would need facilities to keep the weaners separate from the sow (which might mean a great deal of extra fencing to put up at considerable cost). The piglets would also need their own shelter, as well as their own feeding and water troughs. Be very careful not to overstretch yourself. By preparing for the worst, you will not get caught out.

If you are planning on breeding pigs, you should buy the best female you can afford and ensure that you have sufficient land and facilities.

Farrowing equipment

You can buy specialist farrowing equipment, such as an ark with farrowing rails and a built-in crèche, but improvisation is also possible to reduce expenditure.

Farrowing ark or shed

Ideally you should have a suitable place to farrow the sow, such as a farrowing ark or shed. This ark is larger than normal and has steel farrowing rails running around the edges of the ark, positioned about 30 cm (1 foot) in from the sides and the same distance from the floor. Most (if not all) farrowing arks include a floor and ventilation.

If a farrowing ark or shed is not a possibility, a barn or stable will suffice, but you will need to put extra precautions in place to protect the piglets from being squashed by the sow. If it is not feasible to fit farrowing rails, tyres should be hung far enough down the walls to prevent the sow from leaning against the wall when lying down.

Commercial farmers who farm modern breeds intensively tend to use farrowing crates, which follow a similar principle to farrowing rails, stopping the sow from lying on the piglets. These are not popular with breeders of traditional breeds, however, as they are thought to be too restrictive and against the ethics of free-range pig-keeping.

Farrowing crèche

This is useful in that it allows the piglets to wander outside the confines of the ark, yet be contained. The crèche is normally made of steel or wood, and is about

30 cm (1 foot) tall and no wider than the front of the ark. Usually it is attached to the front of the ark, but there are companies that make farrowing crèches an integral part of the farrowing ark.

Creep and creep lamp

An area known as the creep (usually a corner) should be partitioned off from the rest of the pen, to allow the piglets in while keeping the sow out. An infra-red creep lamp should be suspended above the creep to help keep the piglets warm, especially in winter. Heat lamps can be purchased from agricultural merchants, with either a white or red bulb: both do the same job. Make sure the sow cannot come in contact with either the lamp or the flex when you are positioning it, and that you keep a spare bulb in case one blows.

Piglet booster

It is a good idea to keep some piglet booster, which usually comes as a paste that is given orally to weak and distressed piglets. However, it can be hard to source as most vets do not keep it in stock and you may have to contact specialist companies. It can be expensive and has a very short shelf life, so should only be opened when needed and then stored in a cool place.

Surveillance system

If the farrowing sheds are a long way from the house, you may wish to consider installing a surveillance system, although to do this you will also need an electricity supply. Prices vary, but it is possible to buy a reasonably priced system. The camera is set up inside the pen and the monitor is positioned wherever you spend the most time – some people place it next to their bed if they are expecting a pig to farrow during the night.

Iodine

The umbilical cord of each piglet born should be sprayed with iodine as soon as possible, to stop infection entering the cord. Thoroughly spray both the navel and the cord completely. If this is not done, joint ill (see page 74) may occur.

A creep partitioned off from the rest of the pen reduces the risk of the piglets being crushed or attacked by the sow.

Teeth clippers

It is occasionally necessary to clip the piglets' eight eye-teeth (they have four at the top and four at the bottom). These are incredibly sharp and can cause injury to the sow. If it is her first litter, this can lead to her becoming unsettled and she may not let the piglets suck from her. It can also lead to mange in the other piglets, if they are scratched by the eye-teeth in play.

Bucket and spade

The afterbirth can attract foxes, so pick it up and dispose of it as soon as possible. Keep a bucket and spade to hand for this purpose.

Ways of getting your pig in-pig

How you get your pig in-pig depends on a number of factors: your own experience; whether you have a small or large herd; and the facilities you have at your disposal. It also depends on your budget.

Keeping your own boar

This should not be undertaken lightly. Boars need to work and one or two sows will not be sufficient. If not kept properly occupied, over time your boar will end up frustrated, which could lead to behavioural problems. Lack of work could also affect his fertility.

Your stud boar

Many people who take on a boar hire him out for visiting sows for financial gain or if they feel he isn't working hard enough. Don't mate him until he is at least seven months old.

If you are taking in visiting sows, you should put together a brief contract stipulating the fee and what the sow's breeder should expect in the way of care and the service.

When you take in a sow, make sure she is healthy and in good condition, free from any skin diseases and has been wormed recently. If possible, she should come with her own food, so that she does not get a stomach upset from eating different food, which might reduce her chances of becoming pregnant. Remember that once she arrives, you may be on a 21-day standstill (see page 45), depending on your country's regulations.

When considering a boar, bear in mind your own experience. Boars can be unpredictable and should not be kept by a complete novice. Even if they are normally very docile, because of their sheer size they can do damage without meaning to. They need to be kept separate from other males or run with your sows, because two adult males will fight and cause serious damage to each other. Boars are also very adept at escaping if they can smell a female in season, so their enclosure fencing must be well built, with either electric fencing or barbed wire as a back-up.

If you buy a young boar, you must be careful that he is ready for his first service. Try not to mate him until he is at least seven months old and preferably older. He should be up to size and weight – that is, at least on a par with the sow, but not too massive. Make sure that the first gilt he serves is of similar size and of a pleasant nature. If he has problems on his first service, he could be put off for life.

Your boar must of course be fed a nutritious diet – and enough of it. Serving sows or gilts takes a lot out of the boar, so he needs to be fed accordingly. For this reason he should not be allowed to service more than a couple of pigs a week.

If he is allowed to run with the sows, pay careful attention to the service dates and whether the sow returns (comes back into season) after 21 days. However, this is not always possible if you do not live on the premises or if you have to go away for a few days and have a few sows running with the boar. In this case, try to keep a monthly ongoing calendar of the dates when you expect all your sows to come into

Boars are unpredictable and difficult to handle and should not be kept by an inexperienced pig-owner.

In artificial insemination, the boar's semen is introduced into the sow's vulva using a catheter.

season. That way, even if you miss a service, you will have a good idea when each individual sow should return, and you can watch accordingly for this to happen. Records should be kept in any case of service dates, and these should later be completed to include farrowing and weaning dates.

As your boar gets older, you may find that his tusks have to be cut back regularly and this may have to be done by a vet, unless you are experienced at the job. On the whole pigs that are kept outside don't have too strong a smell, but boars can have quite a pungent odour, so you should bear that in mind if you are close to neighbours.

Buying a boar can be quite expensive, so if your budget doesn't run to this, or you do not have the herd size or experience to warrant buying one, there are three other options available to you: artificial insemination, taking your sow to a boar and bringing the boar to you. Bear in mind that should not attempt to breed from your gilt until she is at least 12 months old. If first-time mothers are too young, they will sometimes kill their offspring or not let them suckle.

Artificial insemination

Artificial insemination (AI) is growing in popularity for smallholders who do not, for whatever reason, have access to a boar. Although there is some skill involved, AI is actually a lot easier to perform than people think, and it is certainly a cheaper option than buying a boar or in some cases hiring one. The names of companies offering this service can be found on the internet and in some countries pig associations and rare breeds trusts may be able to help.

The hardest part of doing AI yourself is knowing exactly when the sow or gilt is in season. If you are new to breeding it can be difficult to tell, unless you know what you are looking for. One of the surest ways of telling is if your sow will stand and let you put pressure on her back. Some pigs change their character completely when they are in season: quiet pigs will become noisy, while docile ones will try to escape. Many pig owners just 'know' when their sow is in season, especially if they have had her quite a while and know her character. Depending on the breed, most pigs can reproduce from the age of nine months and come into season every three weeks.

Semen should be ordered by telephone as soon as your pig comes into season. It should arrive in an insulated box by post the next day, although if it is ordered on a Friday, you might not receive it until the Monday, which could of course mean that your pig is no longer in season. Detailed instructions are usually provided with the semen. If the pig is still in season, the catheter can easily be inserted into her vulva and the semen released.

AI gives you access to bloodline and breeding stock, which might not otherwise be available to you. Filling out the paperwork and adhering to the standstill that would have been required if you had brought a boar onto your property is not necessary with AI. And if AI takes successfully, it can be a less expensive way of getting your pig in-pig.

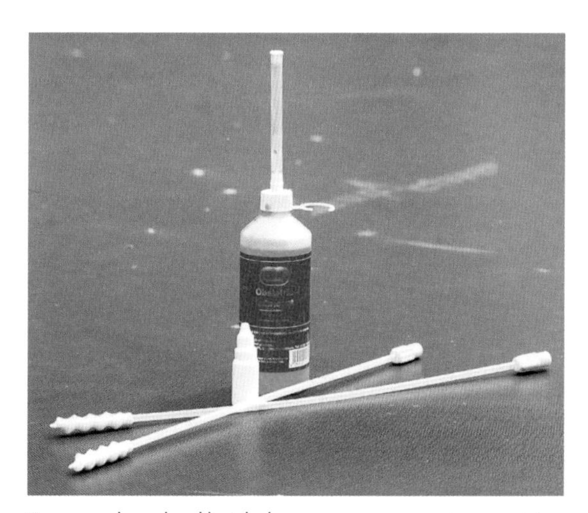

Semen can be ordered by telephone as soon as your pig comes into season and will arrive by post the next day.

Taking your sow to the boar

Some breeders will not let their boar off their property, especially if he is valuable, so you will have to take your sow to the boar. If you want to avoid unnecessary cost, take her just before she is due in season. The breeders will want her to stay for at least two seasons (six weeks) to ensure that she is in-pig, so the closer to the first season you take her, the less it will cost. Don't choose a boar that lives too far away, to avoid stressing the sow by travelling a long distance. If possible, try to visit the boar before arriving with your sow. That way you can check that he is fit and well and has a good temperament. Make sure that your sow is insured before she leaves your property, so you won't be landed with costly vet's bills if something does happen while she is with the boar.

Bringing the boar to you

Only take on the responsibility of having someone else's boar on your property if you have suitable facilities and the experience to handle an unknown boar. Remember that you are running a risk of disease by bringing in a strange pig, so the boar should stay as far away as possible from all other pigs, except the one he will be servicing.

Never put the boar straight in with the sow; instead put him in an adjoining paddock so that both pigs can first introduce themselves through the fence. Try not to just leave them to it when the boar finally does get to service the sow or gilt – take him out of his pen early in the morning, put him in with the sow and watch them. Sometimes one will take a dislike to the other and they will go for each other, so you must be ready to intervene with slapboards if necessary.

As soon as the boar has finished, put him back in his pen and then put him to the pig again later in the day. Do not send the boar back to his owner until you are absolutely sure that your pig has not come back in season. This means that you will have him on your property for at least two seasons (six weeks).

Farrowing

The gestation period for pigs is 116 days or three months, three weeks and three days. Some breeds and first-time mothers don't appear pregnant until the last moment and, if you are unsure about the service date, this can lead to you being unprepared.

Readying the farrowing area

Your sow should be separated from the others at least two weeks before she is due, to give her time to settle into her new home and start nesting, and also to ensure that you are not caught out if you have got your dates wrong. Occasionally, even if you are sure of the exact date, piglets can be born early and catch you unawares.

Before moving your pig into her new home, she should be wormed and the farrowing ark or shed (see page 82) should be thoroughly disinfected and left to dry. Make sure that the piglets won't be able to escape from the ark, by attaching a crèche or a door to the front. Doors should be constructed half of mesh and half of solid wood, so that the inside of the ark isn't too dark once the door is closed. There will then be less likelihood of the sow not noticing the piglets and standing on them.

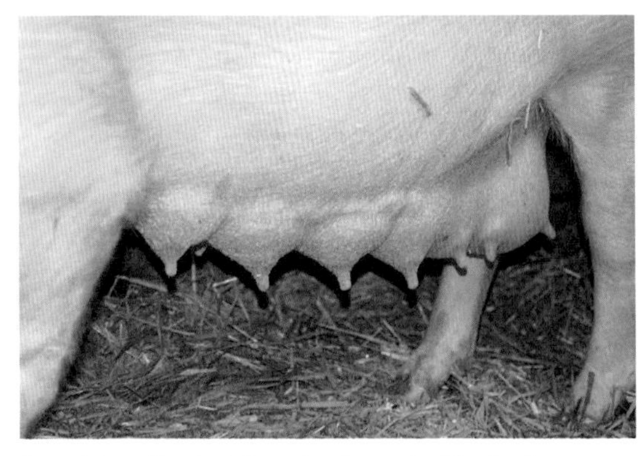

A good sign of imminent farrowing, the sow's udders harden 24 hours before the birth and milk can be squeezed out.

If you are using a shed, you will need to set up the creep area (see page 83), ready for when the piglets are born. Surveillance equipment (if being used) should also be positioned at this time. Limit bedding: too much and the piglets could become buried under the straw and get lain on by the sow. Some breeders prefer to use wood shavings for this reason.

The lead-up to the birth

Once your sow is settled, she should be left to herself as much as possible and allowed to have some peace and quiet. Stress can lead to her aborting the piglets or to stillbirths.

About a month before she is due, she will start to develop an udder. This is always an exciting time, because at last you will be able to see that something is happening. Approximately two or three days before the birth, she will start nesting, pushing straw and whatever else she can find into a heap. She will do this using her front legs and possibly her mouth. Once the nest is to her liking, she will lie down in the middle of it with a contented sigh.

Approximately 24 hours before she is due, the sow's udder will harden and milk can be squeezed out. This is a good indication that the birth is very close.

The birth itself

As the time draws near, the sow's vulva will swell and possibly take on a red appearance. She will become restless, lying down on one side, getting up and then lying down on the other side. When you feel the birth is imminent, wash the pig's udders with animal-safe

Nesting, in which the sow piles up straw in a heap, is a sign that the birth will take place in about two or three days.

disinfectant and make sure that the bedding is as clean as it can be.

In case of problems you should remain close by, but not so close that you are disturbing her. Farrowing can take up to eight hours, and the time between successive piglets being born can sometimes seem quite worrying. However, this is quite normal, so a chair and a flask of coffee are essential for you.

The womb is in two parts, each part of which is called a horn. When giving birth, your sow will lie first on one side and empty one horn, then turn over and lie on the other side and empty that horn – although some pigs alternate between the two. Once she has finished one side, she will expel the afterbirth, which (if you are new to breeding) can catch you out, thinking that she has finished. The second afterbirth should be expelled soon after the last piglet is born. If this doesn't happen, she should be given long-lasting penicillin and some oxytocin.

Pigs do not have a water bag; instead they flick their tail quite violently just before a piglet is born. As each piglet is born, dry it with a towel and put it on a teat to suckle. If a piglet is very weak or not breathing when it is born, rubbing it vigorously with the towel or holding it upside down for a few minutes will sometimes revive it. If it is very cold, you may have to put it somewhere warm in the house, such as the oven on a low heat – but do remember to leave the oven door open!

If you have a first-time mum, she might be confused by the piglets and even attack them. If this happens, they should be taken away from her while she is still 'pigging', put underneath the heat lamp and reintroduced to her when she has finished. If she is still showing signs of restlessness, she should be given a mild sedative. If you suspect that this might happen, because of her nature, it is best to speak to the vet beforehand to ascertain which sedative is safe for her and then keep a small amount handy.

Caring for your piglets

Once all the piglets have been born, make sure that you have disposed of the afterbirth and that all the piglets have had a good feed. Any weak piglets must be shown the teat and if necessary held there while they suckle.

The first hours

Often weak or sickly pigs will come round quickly once they have had the first feed, known as colostrum. This is vital to all newborns because it contains antibodies that help give immunity towards infections. A squirt of piglet booster (see page 83) should also be given to weak piglets. Boosters contain high levels of immunoglobulins and energy content which helps maximizes the benefit that the piglets will receive from the colostrum.

Colostrum is only present in milk for the first 24 hours and starts declining after about six hours, so it is vital that piglets get a good supply within that time. Piglets will drink from the sow every 20 minutes or so, although time between feeds will lengthen as the piglets get older. More often than not they will favour one teat and return to it every time – so much so that udders can look quite strange, with only a few teats in use.

Cold and draughts can kill a piglet quicker than anything, so piglets should be introduced to the creep area (see page 83) as soon as possible. Once they realize they can lie there under a warm heat lamp, they will make their way back to it as soon as they have fed. Lying here also has the advantage of keeping them out of the way of the sow and reduces the possibility that they will be lain on. In cold weather the heat lamp should stay on night and day for at least two weeks, or until you feel the piglets are strong enough to do without it. Do not be tempted to put more straw in to keep them warm – the heat lamp will be sufficient.

If your sow is farrowing in an ark without a door, it is worth putting up a piece of carpet or thick plastic sheeting to help keep it warm. Ideally, the farrowing ark should be insulated to keep a constant temperature inside and it should always have a floor. Do *not* put up a heat lamp inside the ark itself, for there is insufficient room with the sow present and it could be dangerous. Some companies sell heat mats that, once warmed up, stay warm for some hours. If the weather is very cold, consider putting one of these under the bedding.

Some breeders give their piglets an iron injection within a few hours of their birth, but this is not necessary if the piglets are born outdoors and have access to soil. Even if they are born inside, a couple of sods of earth thrown in with the piglets will usually be enough to prevent an iron deficiency.

The first few days

Pay close attention to the piglets over the next few days to ensure they are putting on weight. Even the runt of the litter should appear to have a full stomach, despite being tiny to look at. If you feel the piglets are causing the sow discomfort because of their eye-teeth, it may be necessary to clip them (see page 83). This should be done as early as possible and away from the sow, so that their squealing doesn't upset her.

Breeders who farrow their pigs in arks usually attach a crèche (see page 82) to the front of the ark. This keeps the piglets from wandering off, and the sow can step over it and go off on her own. Care should be taken that very young piglets cannot escape, and that predators cannot take them. Breeders have lost whole litters to foxes, so take precautions if your area has a high fox count or if you live in an urban area where foxes are braver about approaching buildings.

All the piglets should look healthy and well-fed, so keep an eye on them to check they are suckling properly.

Weaning

Piglets should be offered creep food (see page 54) at about three weeks old. If possible, this food should be placed in a hopper and fed ad lib. Make sure it is out of reach of the sow.

At about eight weeks the piglets should be weaned. Two weeks prior to this, the piglet's food should gradually be changed from creep to grower food (see page 54). This will enable the piglets to adapt to the new food before they undergo the stress of weaning. They can be weaned slightly earlier than eight weeks, if it is felt that they are taking their toll on the sow and she is losing condition. This sometimes happens if she has a very large, demanding litter.

When weaning, take the sow away from her piglets rather than the reverse – the piglets will then find the experience less stressful because they will still be in their familiar territory. At the same time worm all the pigs by injection, so that you know each pig has had its full quota.

From around eight weeks, weaners stop taking food from the sow and become dependent on you for their nourishment.

Hand-rearing

Hand-reared piglets will need feeding every two hours – day and night – for the first few days, and must be kept in a warm place. Hand-fed piglets are thought never to do very well and they are usually kept as pets or sent for meat when they are ready. Mostly it is breeders of no more than a handful of pigs who hand-rear; commercial farmers consider it too time-consuming and not cost effective.

It is now possible to get hold of dried sow's milk. Speak to your vet about this before the litter is born in case you need it in a hurry. Finding a substitute for sow's milk can be difficult and you may have to resort to adding glucose to baby milk or condensed milk.

If the piglets have not had any colostrum, you will either have to buy some synthetic colostrum or milk the sow (which is better for the piglets). Even if you don't have any problems with a particular litter, it is worth collecting colostrum and freezing it to give you an emergency supply for future litters.

This Oxford Sandy and Black sow has produced a litter of healthy piglets.

By the time they get to eight weeks most weaners are no longer spending much time with the sow. However, some can get the scours when they are first weaned (see page 72), so they should not go straight to a new home. Hang on to them for about a week after weaning to keep an eye on them. Take all precautions to make sure that your weaners cannot escape – cute though it may seem to see them running towards you, ears flapping. Once they have learned to escape, the weaners will gradually move further and further away from the pens as they gain confidence, which could leave you with all sorts of problems.

The sow should not be put back to the boar straight away, even if she is in reasonable condition. She should be allowed to have her next season, which is usually four or five days after weaning. The season after that is the one when she should go back to the boar, if her condition warrants it.

Once you have started breeding, it is very hard to stop. No matter how experienced you are, you can never fail to be moved by seeing piglets born. Watching them take their first wobbly steps round to the milk bar, and hearing the contented grunts of the sow, delights even the most experienced pig-keeper.

Marketing your piglets

Marketing your piglets and achieving a 100 per cent sale in the required time is a skill that needs to be learnt, but you will get better with practice. A few simple measures will make the whole process easier.

Researching your area

Selling weaners for a good price can be difficult, unless you are well known in the showing world or have an established name for selling good-quality stock. A lot of thought therefore needs to be given to any breeding project, otherwise not only will you become disheartened when you cannot sell your stock, but you could be out of pocket as well.

Before buying a particular breed as your foundation pig, you should thoroughly research which breeds are predominant in your local area and what prices the weaners are being sold for. For example, should you be thinking about breeding Gloucester Old Spots and there is a saturation of breeders in your area, you will be competing against them. Established breeders known for selling good-quality stock can command the highest prices and are always going to capture the market over someone who is little known. So, to give yourself an advantage, choose a breed that is relatively scarce in your area.

Join the relevant breed association, because this is often the first place that prospective owners look when buying a weaner. Get to know as many other breeders as possible, so that they are able to pass on your name if they themselves do not have any weaners available for a customer. It is also a good idea to have a website showing quality photos of your pigs, and this should be kept updated. Inviting the local paper to do a story on your pigs is also a fantastic way of promoting yourself.

Advertising and pricing

The time to start advertising your piglets is once they have been born, while they still have the cute factor. Agricultural merchants may let you put an advert up for

If you are looking for a good price for your piglets, you should avoid selling them at a market or auction.

free and some newspapers also offer free advertising, as do breed societies.

Go through your litter as soon as possible and pick out those you think are suitable for breeding. Breeding pigs command far more money than finishing pigs (pigs suitable for meat). Tag them and make a note of which are males and which are females, and which are for meat and which are breeding stock. You will also need to register the piglets (see pages 44–45). Breeding stock is herd-registered at a later date.

Set a price and stick to it. Prospective buyers can always tell when someone is willing to take less than the advertised price, so price your weaners realistically and don't let yourself be badgered into selling them for less (unless you are selling a few at once, in which case you can discount them slightly). Once word spreads that your weaners were bought for a lesser amount, it will be difficult to sell them for any more.

To give yourself an edge, it is worth advertising that all weaners come with a bag of food and wormed. In this way people will feel they are getting value for money and will be more likely to pay the asking price. Keep all details of the weaners by the telephone, in case anyone phones to enquire about them: pedigree lines, age, the sire's name and dam's name are the sort of information that people may require.

Viewing your piglets

If a prospective customer comes to look at your piglets, the general impression that you create could make or lose a sale. Make sure the pig pens look professional and are as tidy as possible, with hoses, buckets and so on put away. All the pig houses – whether arks or a barn – should have fresh bedding. Troughs should be clean and filled with fresh water, and there should be nothing tied up or fastened with bailer twine. If your customers have come a long way, offer them a drink and a sandwich.

For some people, buying an animal of which they have limited knowledge can be daunting, so reassure them that you are at the end of the telephone, should they need advice. Stand back and let the piglets do the selling. Above all, don't give your customers the impression that you are in a hurry – if they feel uncomfortable with you, they won't buy.

Once a sale has been agreed, take a deposit and don't be caught out by the new owners asking you to keep the piglets for an extra week or so because their fencing isn't ready or they are going away on holiday. Keeping them for longer than planned will eat into your profit, so make sure you charge accordingly.

Selling will get easier with every litter, until you have reached the point where you have a waiting list. Concentrate on building up your name and breeding good-quality registered stock, and it won't be long before you are known as an established breeder.

You have invested dedication, time and money into raising your pigs so don't undervalue them when it comes to making a sale.

Processing

Provided your animals are processed humanely, the enjoyment of eating your own home-produced meat is second to none. This chapter explains the slaughtering and butchering processes, together with the rules and regulations that accompany them.

Preparing for slaughter

Nothing tastes nicer, or gives more satisfaction, than sitting down to your own produced pork. Traditional breeds have to be eaten to survive, and as more people become aware of their eating qualities, these pigs are slowly making a comeback.

Traditional breeds

Raising traditional breeds for meat is a slower and more costly process than rearing intensively produced pigs; this means they suit the outdoor system far better than leaner, more modern pigs such as the Landrace (see pages 120–123). Over the years, as tastes changed and the fattier meat associated with traditional breeds fell out of fashion, these breeds declined to such as extent that most came perilously close to extinction, although they are now undergoing a revival in popularity.

It is the slow growth and the extra fat associated with traditional breeds that makes the meat of these pigs taste as good as it does. That, along with exposure to the elements, exercise and a generally stress-free, more natural life, gives the pork a succulence that is hard to match.

The run-up

Both male and female pigs are good enough to eat. Many people (including butchers) will often refuse to buy meat that has come from a boar, due to what is known as 'boar taint' – a very gamey smell and taste. However, this is rare in outdoor pigs – especially traditional breeds – and should not be the reason for buying solely female pigs for meat.

When raising pigs that are intended for the freezer, it can be difficult to keep them in good condition, without letting them get too fat. Keep a close eye on how much food is being given. Most pork pigs go for slaughter at around six months, but if you have a breed that is prone to a lot of back fat, this may be a bit too long. Some breeds, such as the Saddleback (see pages 128–131), get to a stage where all they do is start putting on fat.

An abattoir should be chosen and the pigs booked in at least a month before they are due for slaughter. Ideally the abattoir should be a family-run local one. Many smallholders find that the smaller the abattoir, the more willing the staff are to offer help and advice. It is also thought that pigs do not suffer as much stress at a small abattoir as they do at a large one. Some abattoirs do not take pigs and, if they do, often have a specific day for slaughtering them, which can result in quite a long wait before they have a space for you.

If you want crackling the skin has to be left on, so make sure, before you take your pigs in, that the abattoir's processing methods leave the skin intact. This requires special, expensive equipment, which many of the small abattoirs do not have.

Freezer space should be given due consideration prior to the pigs' slaughter. Do not underestimate the amount of meat that you will get off the pigs. If you are taking more than one pig to the abattoir, you could find that you do not have the space to hold all the meat. If in doubt, you may have to buy another freezer.

Pig welfare

It is vital to keep stress to a minimum. Pigs that are destined for the abattoir should be able to walk into the trailer quite happily. If this is not the case, start training them two or three weeks beforehand. Stress can affect the quality of the meat, so everything should be done to keep the pigs calm and unperturbed.

On slaughter day it helps if you have trained your pigs from a young age not to be afraid of the trailer.

On the day that the pigs are going for slaughter, make sure the trailer is well bedded down and do not overcrowd it. If you are transporting a few pigs to the abattoir, it is better to take them on two separate occasions rather than cramming them all in together. Make sure that the pigs are as clean as possible as the abattoir might reject them if they are dirty. Do not leave them in the trailer for too long prior to your departure and, if you can, time the journey so that you arrive just before they are due to be slaughtered.

Ensure that you are able to reverse your trailer. Bear in mind that you may have to reverse quite a long way and there may be obstacles, such as other cars, in your way. If you are at all unsure about your reversing skills, practise without the pigs on board first.

If your pigs have been on any medicine, be sure that the withdrawal period has ended. You will need to sign a form confirming this.

Many pig owners are worried that they will not be able to take the pigs to the abattoir when the time comes. It will be hard the first time you take them, but it does get easier with each pig that you take. You may suffer less guilt if you don't name them, because this sometimes helps you stay detached from them.

Weights

There is no hard-and-fast rule concerning the weight at which you send your pigs to the abattoir. It very much depends on how large or small you want your carcass to be and for what purpose it is intended.

Formula for measuring a pig

If you have not yet developed an 'eye' for your pigs' weight, you can buy a weighing tape. Alternatively, use a tape measure that is long enough to reach right around the pig's body and make a simple calculation as follows, measuring in either inches or metres:

- Measure right around the pig just behind the front legs and square the result.
- Measure from the base of your pig's ears to the base of its tail.
- Multiply the first figure by the second. Then, if using metres, multiply the result by 69.3 to obtain a weight in kg; if using inches, divide the result by 400 to obtain a weight in lb.

There are various methods of calculating your pig's weight without scales, using a weighing tape, a tape measure or just your 'eye'.

Meat categories

These stages of production are not set in stone, although most owners send off their pigs at the porker stage and often ask the butcher for bacon and sausages, as well as the more usual joints.

- **Suckling pig** Usually a very young pig still on the sow, weighing around 15 kg (33 lb); popular with the restaurant trade.
- **Weaner** A newly weaned pig, usually between eight and ten weeks old and weighing around 20 kg (45 lb); often sold to smallholders who wish to fatten the pig themselves or sent on to finishing units to fatten for the commercial trade.
- **Porker** A pig around five to six months old and weighing approximately 55 kg (120 lb).
- **Cutter** Kept for longer than a pork pig, usually up to about nine months, and weighing around 65 kg (145 lb).
- **Baconer** A pig grown on specifically for bacon; it tends to be between nine and 12 months old and weighs around 80 kg (175 lb).
- **Sausage pig** Pigs used specifically for sausages tend to be older sows or pigs that are no longer required; finding an abattoir to take such a large animal can be difficult.

Paperwork

Different countries have different paperwork requirements and it is important that you conform to these. In the UK the current regulations require a food chain information form to be filled in to accompany the pigs when they go for slaughter.

Essential information

EU regulations that came into force in 2006 recognized all primary producers as food producers, so even if you are keeping the pigs solely for your own consumption, you will be required to fill in the food chain information form. This form can be obtained from the abattoir or downloaded from the BPEX (British Pig Executive) website. BPEX also has an online system that enables you to send the required information in advance to abattoirs, as long as it is with them before 4 a.m. on the day the pigs are due to go. The information is then relayed to the abattoir in an email that is sent at 5.30 a.m. It is also permissible to design your own form, although this has to be approved by the abattoir before you take in your pigs.

All three sections of the form should be filled in and signed. The following information is required:

- **Producer copy** The name and address of the producer and the identification of the pigs going to slaughter. This copy should be sent to the relevant local authority department within three days.
- **Haulier copy** Details of the haulier and identification of the pigs being transported.
- **Processor copy** The processor's name and address, and any information about medical products or other treatments administered to the pigs during the last 28 days. The abattoir keeps a copy of this for their own records.

If you are travelling more than 65 km (40 miles) to the abattoir, you will also be required to gain your Certificate of Competence by undergoing a multiple-choice exam that tests your knowledge of the welfare of the animals while travelling, current transport regulations, loading and unloading and what to do in an emergency or vehicle breakdown. This certificate is required if you will ever be travelling more than 65 km (40 miles) with pigs, for whatever reason.

Always keep a file copy of any paperwork that is associated with your pigs so that, in the event of being inspected, you have the records to hand. Relevant travel documents, such as the Certificate of Competence, should be carried with you at all times when transporting animals.

Having cared for their pig throughout its life, owners are relieved that the abattoir treats the animals gently and humanely.

What to expect

Most pig owners have never stepped inside an abattoir or seen their pigs killed. They offload their animals and then leave straight away, either because they are squeamish or because they feel guilty.

Preparation

Most family-run abattoirs treat the animals as gently as possible and with consideration. The majority of abattoirs are clean and quiet, with no obvious signs of animals waiting around. There is usually a strong smell of disinfectant, rather than of death.

Before leaving for the abattoir, check that you have the right paperwork, filled out correctly, as this will have to be handed in before offloading the animals. If the abattoir is also butchering your pig, you should give them a written list of what is required, or you may find that you don't receive what you asked for.

Once you have booked in the pigs, you will be given a time at which to drop them off, usually early in the morning. On arrival, a veterinary inspector will check that the animals are fit and well, with no obvious defects, and could reject them if they are deemed too dirty.

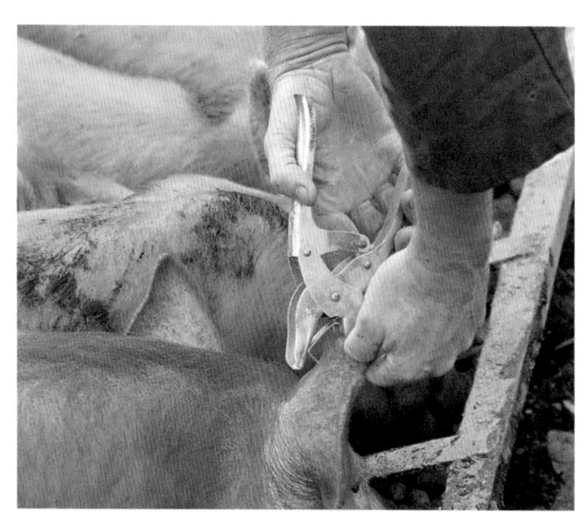

Before being delivered to the abattoir, pigs should be identified with a metal tag recording your herd number or by slapmarking.

The slaughter process

Pigs are stunned with an electrical current from tongs applied to each side of the head. The pig is rendered unconscious and is shackled up by one hind leg. Once it is raised clear off the ground, all the major blood vessels that leave the heart are cut, which produces rapid blood loss. It is this bleeding that actually kills the animal. The whole pig is then dipped in a tank of near-boiling water to loosen the bristles, before they are removed by the paddles of a dehairing machine. Once the carcass has been split and evisceration or disembowelment has been carried out, it is checked again by a vet, before being washed and chilled and then hung in a cold room awaiting collection.

The whole process from arrival at the abattoir to slaughter is carried out as quickly as possible. No one knows for sure if pigs are aware of their fate, but as long as they have had a good natural life, are stress-free on their final journey and handled in a humane way in their final hour, you will be able to load your pigs for the last time with a clear conscience.

You will need to wash and disinfect the trailer within 24 hours of leaving the abattoir. Some abattoirs require you to do this before leaving their premises and will show you the washing area.

If you are collecting the carcass to take to a butcher, you will be told when to pick it up. This is normally two or three days later.

Butchery

Once you have a slaughter date, you should start looking for a butcher to process your carcass. Many butchers get booked up quickly, so make sure the one you choose can fit you in on the day your meat is going to be collected.

Finding a butcher

Some abattoirs have their own butcher on site, which will save you time and money, since you don't need to go back to the abattoir to collect the carcass. However, most abattoirs offer only a basic joint-cutting service, and perhaps sausages as well, so you could find that your choice of cuts is limited.

Choosing a butcher outside the abattoir is not without its difficulties, too. Many first-time pig owners do not really understand the cutting process or what to ask for, so try to choose a butcher who has been recommended to you by another pig-keeper and with whom you feel you can get on. Not all butchers have the equipment needed to smoke or dry-cure meat, so before you book in the pig make sure that the butcher you choose is able to carry out all your requirements.

Once the pig has been killed, meat needs to set, otherwise the butcher will have problems cutting it. Unlike cattle, pigs require only two or three days to set, although they have been known to be hung for longer without problems. If the meat has been collected too early from the abattoir, it might not have set, which will result in 'floppy' pork. Take advice from your abattoir on the collection date.

Someone at the abattoir or butcher will be able to give you an idea about whether you were on the right track with the fat content of your pig. They will let you know whether the back fat was too thick or just right. This will help you to feed your pigs accordingly the next time you are raising them. The butcher may talk about the killing-out percentage and the lean-meat percentage. The killing-out percentage is the proportion of the live weight of the pig that the carcass represents, and includes the bones, head, skin and so on. The lean-meat percentage is the meat you actually get from the carcass.

Home butchering

If butchering at home appeals to you, you may have to register with your local authority as a food premises, depending on what you are planning to do with the cuts of meat. Once you register you will be subject to inspections by the local authority and it is your responsibility to adhere to all the regulations. Seek advice before going down this route.

Before trying to butcher a carcass, get some detailed information about the different cuts and which part of the body they come from. Speak to a helpful butcher or buy a DVD or book on the subject (it has been known for pig owners butchering meat for the first time to have a go while watching a DVD on the subject, but this is not recommended). Ideally, find out if a butchery course is being held in your area and then book yourself on it. Butchery is a skilled job and while you may learn the basics from a book, you need hands-on experience to be able to do it reasonably well. Even if you are not planning to butcher yourself, you need to recognize the different cuts and which part of the pig's body they come from so that you know what you are getting back from the butcher (see page 104).

Once you have mastered the relevant skills, there is no end to what you can do with your home-produced pork – whether you are smoking or curing it or simply making sausages.

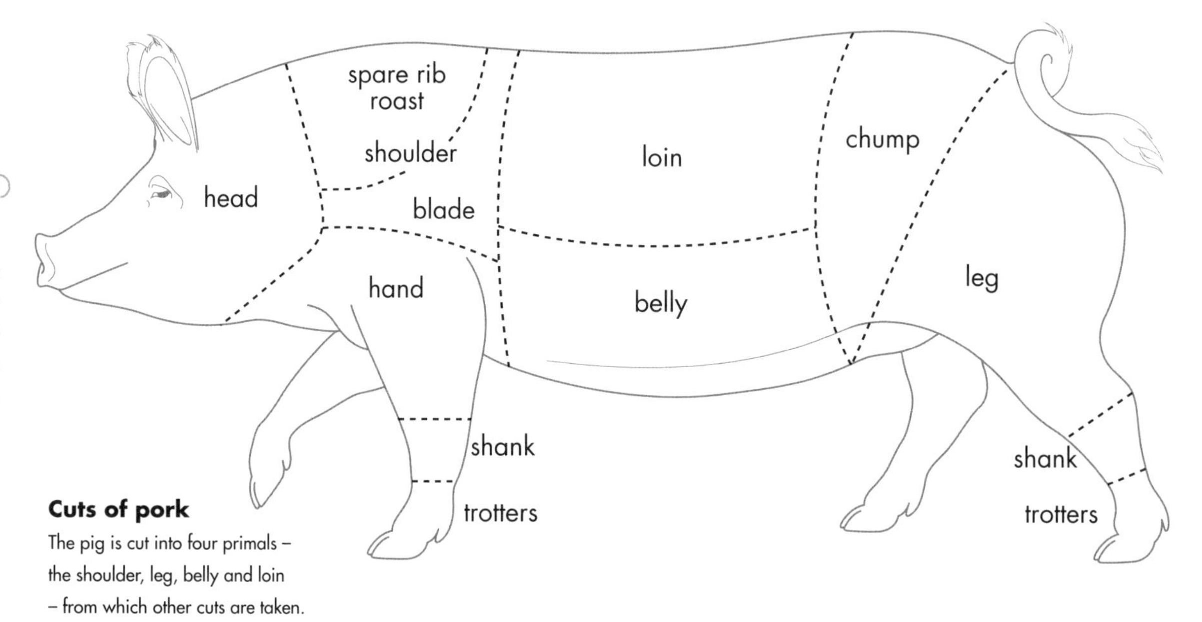

Cuts of pork

The pig is cut into four primals – the shoulder, leg, belly and loin – from which other cuts are taken.

Meat cuts

From tail to trotters, there is not a single part of a pig that cannot be utilized. The head can make soup, and even the ears can be fried or baked. Each side is cut into four primals: the shoulder, belly, loin and leg. These can be further cut or processed as follows:

- **Shoulder** Joints (either boned or boned-out), spare rib and steaks.
- **Leg** Steaks and joints (boned or boned-out); the ham comes from the back leg.
- **Belly** A fattier meat that is used for steaks or diced stir-fry meat; it may also be cut for streaky bacon.
- **Loin** This can be deboned and rolled (rolled loin); pork chops are also usually taken from this area.

Sausage lore

The history of the sausage began in Sumeria 5,000 years ago. Brought to the UK by the Romans, the sausage's name derives from the Latin *salsisium*, which means salted. In the fourth century Emperor Constantine banned sausage-eating because of its link to a pagan festival.

Today, countless recipes exist all over the world and butchers continue to create sausages with new distinctive flavours. Sausages are often named after the area in which they are created, such as the Gloucester sausage, a sausage using the pork from the Gloucester Old Spot.

Pigs on Show

Showing the pigs you have bred and reared is highly rewarding, both in terms of recognition for the work you have put in and for the new contacts and experiences you will gain. A few guidelines on show etiquette and ethical behaviour help to make everything run smoothly.

Showing

Nothing beats the pleasure of winning a rosette with a pig that you have bred yourself; the rewards and opportunities that follow are numerous and range from marketing benefits to finding potential customers.

Why show?

There is no definite answer to this question. Breeders show for all sorts of different reasons: perhaps they like the glory of winning, or they see the advantages of using the show ring as a marketing tool, or they simply enjoy the social side of showing. Whatever the reason, you'll find that the rewards and opportunities that can be opened up by showing your pigs and winning are immense.

Comparing pigs

Showing your stock in the show ring is an ideal way of comparing your pigs with the competition and seeing what other breeders are rearing. If you are new to showing, it is worth getting other breeders' opinions, as well as asking the judge at the end of the class what he thinks of your pig. Most will give you constructive criticism and often they will suggest bloodlines to try if you wish to improve your herd or lose a conformation fault that seems to keep reappearing in your litters. Don't be too disheartened if a judge does not like your pig – different judges have their own particular preferences as to what they do or don't like in a pig.

For many breeders, being rewarded with a first prize in the show ring is the result of years of hard work, possibly initial wrong choices in bloodlines and often lots of disappointments. It is rare for a breeder's first litter to have a potential champion among the piglets and, even if it does, a piglet that could be destined for the top often goes unnoticed through lack of experience.

The best part of showing any animal is often the preparation – and many a prize has been won or lost on the strength of the preparation. A pig won't win, even if it is the most perfect pig that has ever lived, if it is not well prepared and in excellent condition. Judging it so that the pig is at its absolute prime on the day can be difficult and needs to be practised; it is a challenge for breeders, no matter how much experience they have, to get it just right.

A marketing tool

Showing is the best marketing tool there is and full advantage should be taken of it. It can work against you as well, so it is important that your pig is as ready as possible for the ring. Prospective customers like to see ribbons hanging above the pens, as it gives them confidence that they will be buying stock from a quality breeder. Throwing away a placing through lack of correct preparation therefore represents a wasted marketing opportunity. And going home with no rosettes is not only disappointing, but could affect the sale of future litters. When selling to the public, it is a fact of life that breeders whose stock regularly takes home prizes have a distinct advantage over those whose animals rarely get placed.

Promotion opportunities don't end when the show ends. If you have picked up a rosette, your name will often appear in the relevant farming and specialist newspapers, especially if you have won a championship. More often than not potential customers will take note of your name and will come to you in future for weaners.

Meeting customers

Whenever prospective pig-keepers are being advised on where to see pigs and meet breeders, agricultural shows are always mentioned. This is because they enable the potential pig-keeper to view hundreds of pigs of all ages and breeds. It gives them a chance to talk to the breeders about their ideas and get some feedback. For the breeders, this opportunity for speaking to hundreds of potential customers under one roof should not be wasted. Be sure to meet everyone with a smile and, if it is an inopportune moment to talk, arrange a time later on that is convenient to both of you.

Networking

If you are new to breeding and showing, there is no better way of getting to know other breeders than by speaking to them in the pig pens. An even better way of introducing yourself to other breeders at shows is over a pint of beer in the pig tent. Many of the large shows have a refreshment tent solely for the breeders, and sometimes events such as barbecues and quizzes are put on in the evenings. Most show people are friendly and will be happy to talk to you and help, if they can. Whatever reason you have for showing, the day should be enjoyable for both you and your pigs.

Winning a prize at a show is one of the best ways of generating publicity about you and your pigs.

Preparing to show

To get placed in a class, a pig has to have good conformation and be as close as possible to the breed standard. But picking a winner is a skill that only comes with lots of experience.

Show standards

To stand a chance of the top prize, a pig has to have something extra, that 'look at me' attitude and a presence that will catch the judge's eye as soon as it walks into the ring. A pig with good conformation, but without that little bit extra, will probably be placed, but first prize will always elude it.

Before taking your own pig to a show, it is worth visiting as many agricultural shows as possible to get some idea of what the judges are looking for and how the pigs are presented in front of the judge. Choose the relevant breed class and study the pigs in all the classes very carefully as they walk around. Pick out the pigs that you think are going to come first, second, third and fourth. If you're wrong, speak to the judge when judging has finished about why those particular pigs were placed. It is also worth asking the owner of the winning pig if he or she would mind you taking a few photographs, so that you can compare them back home with your own prospective show pig.

Try to be around the pig pens as the animals are being prepared. Most breeders won't mind if you stand and watch them wash and get their pigs ready for the show ring – you may even glean a few showing tips.

Selecting your show pig

When you are selecting your potential show pig, you should have a clear idea of which classes you wish to enter, so you need to plan a few months ahead. Start by contacting the secretaries of the shows in which you would like to participate and they will send you a schedule with a list of the classes. If you have no idea which shows hold pig classes, contact the relevant breed society as they will be able to tell you which shows are coming up or which websites to visit to find out. Some specialist magazines also print a list of shows.

Make sure your pig is as clean as possible before taking it into the show ring.

Black pigs are required to be oiled for the show ring. Keep stress levels down by doing this in good time before the class.

At most of the large shows, classes are split into male and female, January-born, July-born and September-born. There may also be mature classes and in-pig classes, depending on how large the show is. Many shows nowadays also include agility and handling classes for children. Obviously it is important that any pigs taking part with children are of a calm nature and can be trusted to behave themselves, as far as possible.

Agility classes, in which pigs are guided around traffic cones and other obstacles, are great fun. These novelty classes tend to be held once serious showing has ended and pigs that have shown throughout the day are usually entered. It is rare that a breeder will bring a specific pig just for the agility class.

Once you have chosen the class, you can select the pig that meets the criteria required to enter. Be honest about the pig's age: a pig that was born in late December should under no circumstances be put in the January-born class. If you can, try to enter both a younger pig and an older one, in order to give yourself a chance at more than one class.

When selecting pigs to show, make sure you know the breed standard. If you are inexperienced, keep a reference for the breed standard as well as photos of a pig that is true to type in front of you as you make your selection. Go through the conformation points slowly and tick those you think your pig has. The pig that ticks the most boxes is the one to take. All pigs destined for the show ring should at least be straight, level movers, their underlines should be correct, their backs should be level and they should have a good deep body.

The big day

Even if you are incredibly experienced, taking part in a show is nerve-racking, so it pays to be well prepared. The day before, plan your route, pack all necessary equipment and prepare the trailer with enough straw to keep the pig clean. If you anticipate problems loading your pig, do not feed it the night before so that you can use its feed to bribe it onboard. Arrive at the show at least two hours before the start of your class to enable you to wash your pig if required. Even if you bathed it a couple of days before, you may need to freshen it up. Be ready to enter the ring at least ten minutes before the start of your class. Avoid doing any oiling or any final grooming of your pig or yourself in a rush at the last minute, as this will stress both you and your pig.

Rules of showing

There is an unwritten code of conduct that you need to follow when you are showing your pig if you are to get the most out of the situation and keep on good terms with all your fellow pig breeders.

Ethics of showing

Most pig exhibitors know one another very well and this can be a problem for judges. If the situation is not handled carefully, it can lead to all sorts of rumours. If you are friends with one of the judges, you need to make a point of keeping your distance at the show and preferably for the week leading up to it – don't compromise the judge in question.

Most judges stay away from the pig pens on the day of judging and turn up at the pig classes only a few minutes before they are due to judge. They should have no idea which pigs are entered and who they belong to, because they should not have seen the show catalogue.

Behaviour checklist

- ✔ Don't compromise the judge or create a conflict of interest.
- ✔ Be polite but concise in your comments.
- ✔ Present yourself and your pig immaculately.
- ✔ Don't cut up other exhibitors' pigs.
- ✔ Don't use any advertising, unless it is the sponsor's official advertising.
- ✔ Accept the judge's decision with good grace.

Behaviour in the ring

It is usual to say good morning or good afternoon as you enter the ring and to answer any questions that may be put to you by the judge or steward as concisely as possible. Do not try to engage the judge in conversation, unless he or she encourages it – and even then, keep it brief. You should keep your pig under control at all times.

Both you and your pig should appear clean and tidy and well presented. Your slapboard and stick should be free from dirt, with no signs of advertising on them (although sometimes you will be given a slapboard to take into the show ring that advertises the event's sponsor).

Walk around the ring in a clockwise direction, keeping the pig between you and the judge. Never cut up another pig if the judge is looking at it. Hold your board in your left hand and the stick in your right, so that you are not obscuring the judge's view of the pig. Avoid chatting with the other exhibitors and instead concentrate on keeping your pig moving calmly and at an even pace, while keeping a constant eye on the judge. When you know he or she is looking, make an extra effort to present your pig in a manner that shows off its good points. You must stop your pig as soon as you are asked, to enable the judge to bend down and check the underline. Try to keep the pig as still as possible until the judge has made it clear he or she is ready to move on to the next pig.

Make sure you are as well presented in the ring as your pig, with clean overalls, boots and slapboard.

Avoiding conflicts of interest

If you discover that your pig was bred by the judge, you should make this clear to the steward before the judging begins. You will then probably be asked to stand aside until he or she has judged the other pigs, then another judge will then be called in and asked to place your pig in relation to the other ones. If possible, you should try to avoid entering classes that you know will be judged by someone who has had any sort of interest in your pig.

If you lose, accept the judge's decision with good grace. Never argue with the decision – all judges have personal preferences. The judge should come to everyone in the ring in turn and give them his or her comments. Even if you don't agree with the comments, thank the judge. There is always another show.

Waiting to leave

As an exhibitor, you should not leave the showground until the allotted time when the show is officially over. While you are waiting, you should ensure that your pig is comfortable and has food and water. Showing is hard work for both of you.

If you know you will have quite a wait before being allowed home, you can make positive use of the time once you have finished showing by meeting all the people who have tried to talk to you earlier in the day about your pigs.

Agility classes are great fun and offer a chance for competitors to unwind at the end of the day.

Make sure you fill out all the relevant paperwork before you leave the show – and don't forget to send it off to the necessary department once you get home.

Leaving the show

Once the time has come for exhibitors to leave, there is usually a mad dash when everyone tries to bring their trailers to the pens. If you don't feel confident about your reversing skills, it may be worth waiting until a few of the exhibitors have left and you have more room for manoeuvre.

Load your kit first, so that your pig is not left hanging around too long in the trailer. If you see that someone else could do with assistance in loading a stubborn or awkward pig, give them a hand: you might be in the same position yourself in a few minutes' time.

When loading your pig, drive the trailer as close to your pen as possible. If you are anticipating problems, find some portable gates to make a short walkway or lots of people with slapboards.

Back at home

Once home, the pig should be placed in an approved isolation unit, if possible. If you don't have one, be aware that you are on standstill for 21 days (see page 45). This means that pigs can come onto your property but not leave it, unless you are taking them straight to the slaughterhouse.

Your pig should be made comfortable and given food and water. Within 24 hours of arriving home, your trailer should be cleaned out and disinfected so that it is ready to use again.

Keep a watchful eye on your pig for the next few days in case it has picked up an infection at the show. Just in case, it is worth carrying out the same bio-security measures as you would for a sick pig. If possible, keep it isolated from other pigs, wash your hands and use a footbath before and after handling it and wash any equipment carefully until you are sure that the pig has not been infected (see page 65).

Join the club

If you are a first-time pig breeder, you may feel unsure initially about the correct way to do things. Most people muddle through, but this need not be the case. There are many ways of accessing information and one of them is to join the breed society.

Breed clubs

Most of the traditional breeds have a breed club. You join for a small fee and two or three times a year receive a newsletter packed with useful information. You will also be able to access the members' area of their website, which is especially useful if you have stock for sale. More importantly, you will be put in touch with other breeders of your specific breed, some of whom may live close by. Without the breed society, you might never have found out about them. After a while you may even wish to sit on the committee and play an active role in running the club.

Getting to know other breeders is important, because not only will you be able to telephone them if you need any advice, but the other breeders will also become aware of your business and may put customers your way. There is usually a good social life as well, with quizzes, days out and workshops often being arranged. You will be able to improve your stock, as you will be informed about boars that are currently at stud. And, if you are unsure, you will also be able to get advice on which of your subsequent weaners are suitable for showing or breeding.

Using the internet

There are numerous internet forums that you can access, some of which are independent, while others are attached to specialist magazines. All of them are free: you just have to enter your email address and a password and you can use the site and start posting questions straight away.

The great thing about internet forums is that you will get answers to your questions from smallholders and farmers all over the country and sometimes from even further afield. And because the answers come from people in many different situations, you will gain a much better perspective than you would do from talking just to one person.

Many places run smallholders' clubs. These are not specifically for pig owners, but for anyone and everything connected with smallholdings. Depending on which club you join, most (if not all) will run regular courses in everything from fencing to making yogurt. If there are currently no pig courses, why don't you suggest running one?

Whatever your breed of pig there will be a club you can join to gain the advice and support of your fellow owners.

Pig Breeds

This directory takes a look at all the main pig breeds, from the easy-going Landrace to the mischievous Tamworth to the intelligent Vietnamese Pot Belly. Each breed is described in terms of its appearance, personality, preferred habitat, management needs and breeding ability, and a brief history of each pig explores its origins and how it has developed over the years. The breeds are grouped broadly by region, starting with those developed in Britain and continental Europe and moving through the American breeds to those from Asia and the Pacific area.

Large White

hardy • prick ears • known for improving other breeds • large litters

TYPE: rugged, modern breed, with well-muscled hams • SIZE: large • COLOUR: white

The Large White can be traced back to the Yorkshire pig (see page 166). The breed was first recognized in 1868 and the herd book was published in 1884. It was one of the original breeds of the National Pig Breeders' Association, which today is known as the British Pig Association (BPA).

The Large White was originally developed as an outdoor breed, but it was found to do equally well when kept in intensive production systems. Nearly all of the cross-breeding programmes that have been undertaken using two or more breeds involve either the Yorkshire or the Large White, as both these breeds have the ability to stamp uniformity and quality on piglets from almost any other breed.

In the early 1970s there was an increase in worldwide demand for the Large White, and from 1970 to 1973 more than 8,500 Large Whites were exported from the UK all over the world.

Looks

The Large White has fine white hair, a long body with excellent hams and a slightly dished face (upturned nose). It has distinctive erect ears and a dished face.

Personality

This is an energetic but docile pig, with a good temperament. It is easy to keep.

Habitat

The Large White is popular with commercial pig farmers and therefore tends to be farmed in indoor systems. However, it is known for its ability to thrive in any climate and would also do well in a free-range environment.

Management

The Large White is popular with commercial pig farmers, who often use it to improve other breeds.

Breeding

Large White sows generally have very large litters of 12–14 piglets and are capable of producing plenty of milk for them.

Large White

One of the largest breeds of pigs, the Large White nonetheless has a docile temperament.

Landrace

easy-going • lean meat • pork and bacon pig • large litters

TYPE: versatile, modern breed suitable for indoor management • SIZE: large • COLOUR: white

The name Landrace derives from the German term for 'national breed', and today the United States as well as most European countries have their own Landrace breeds. In countries such as Denmark the Landrace is the dominant breed and indeed at times it has been the only one. In the United States, the Landrace is today the fourth most-recorded breed of swine and it is also one of the most popular breeds with commercial producers in the UK.

The Landrace was first registered in Denmark in 1896. In 1934, the Department of Agriculture in the United States imported its first Danish Landrace, originally only for use in cross-breeding programmes to create new breeds. However, in the late 1940s, Denmark allowed the United States to develop a pure-breed American Landrace. A breed society was formed in the 1950s and this association joined the National Swine Registry in 1994.

In 1949, the first Landrace pigs – four boars and eight gilts – were imported into Britain from Sweden. From 1953 onwards, Northern Ireland, the Isle of Man and the Channel Islands also welcomed their first Landrace pigs.

The British breed society was formed in 1949 and the herd book was established with the first piglets from the 1949 importation, which were born in 1950. When Denmark started to produce Landrace for the English market,

British Landrace

Its light-coloured skin makes the British Landrace susceptible to sunburn.

French Landrace

they altered the conformation of the breed by crossing it with a Large White (see pages 118–119) to create a pig that would produce a lean and long carcass.

By joining with the National Pig Breeders Association in 1978 (now the British Pig Association), the British Landrace Society contributed to the development of the pedigree-pig industry and the establishment of a national herd book for all British breeds.

In the 1980s, in order to keep the Landrace breed strong, new bloodlines were imported into Britain from Norway. Landrace pigs in Northern Ireland have benefited from bloodlines imported from Finland and, more recently, from Norway.

The bloodlines were carefully chosen to broaden the genetic base of the pigs, as Landrace breeders are always striving to meet the ever-changing commercial expectations put on the breed. This work continues up to the present day.

Looks

Most Landrace pigs have similar characteristics, although some are of a slightly heavier build than others. The Landrace is a long, lean pig with lop ears. Its head should be light in colour and of medium length with a medium-sized neck. Its back should be long and slightly arched, with full hams rounded from both the back and the side. Its hair should be fine and white.

Personality

This is a docile and easy-going breed, which makes it ideal for intensive management.

Habitat

Indoor rearing may suit thin-skinned pale Landrace pigs more than a free-range environment. This breed is usually found in commercial indoor systems.

Management

The Landrace has been specially bred for speed of growth. Its greatest strength is its ability to improve other breeds of pig when crossed in order to produce hybrid gilts. More than 90 per cent of hybrid gilts reared in western Europe and north America use Landrace bloodlines as the foundation for the profitable production of quality meat.

Breeding

Landrace sows have the ability to produce and rear large litters of 12–15 piglets, with very good daily weight gain.

American Landrace

Oxford Sandy and Black

attractive • hardy • multi-purpose pig • natural forager

TYPE: reasonably lean, traditional breed • SIZE: large • COLOUR: sandy, marked with random black blotches rather than spots, and with pale feet, blaze and tassel

One of Britain's longest-surviving breeds, the Oxford Sandy and Black has been farmed since the eighteenth century. As its name implies, it was particularly popular in the Oxfordshire area – it is also known as the 'Oxford Forest Pig' as well as 'Plum Pudding'. The breed is thought to be related to the old Berkshire (see page 132) and the Tamworth (see pages 138–141).

Despite its earlier popularity, numbers have declined so much that the breed faced extinction on at least two occasions. Breeders worked hard to ensure the survival of the breed, even though it did not have its own society or herd book. It was hoped the Rare Breeds Survival Trust would recognize the breed when the trust formed in 1973. This was not to be, however, and it was left to the small band of breeders to keep the pig going.

In 1985 the current breed society was formed, which helped stem the breed's decline. The first herd

Oxford Sandy and Black

book lists 29 herds (15 boars and 62 sows). Although some bloodlines have since been lost, the Oxford Sandy and Black is now gaining ground with even the rarest of bloodlines improving.

Looks
The Oxford Sandy and Black is a medium to large lop-eared pig. It has a moderately long head with a slightly dished (upturned) muzzle. Its body is long and deep, with broad hindquarters. Its legs are of medium length and well set.

Personality
This is a docile, friendly pig that likes to be with people. It is a good choice for beginners.

Habitat
This pig enjoys woodland or pasture environments. A hardy breed, it is suitable for all climates.

Management
Oxford Sandy and Blacks are economical to keep as they are natural browsers and foragers. Less inclined than some other breeds to put on excess fat, they are ideal for people who are new to pig finishing. Their colouring means sunburn is not a problem.

Breeding
Excellent mothers with plenty of milk, the sows farrow around six to nine piglets.

Although large, the Oxford Sandy and Black is an easy pig to manage.

Gloucester Old Spot

docile and hardy • lop-eared • quality meat • suitable for novices

TYPE: traditional, true smallholder pig that produces the finest-quality meat • SIZE: large • COLOUR: pale whitish coat with random black spots

It is thought that the Gloucester Old Spot evolved from crossing an old-type Berkshire (see page 132) with an original Gloucestershire pig. It has been around for two or three centuries, but was only officially recognized in 1913. It is known as the 'Orchard Pig' because it was often kept in apple orchards in Berkeley Vale, Gloucestershire. Its spots were reputedly caused by apples falling and bruising its skin, and over the past decade the acceptable number of spots visible on the breed has changed from just one or two to several.

As specialist markets for traditional free-range breeds grow, the meat of the Gloucester Old Spot is increasingly in demand. Today it is the largest numerically of the pig breeds listed by the Rare Breeds Survival Trust and is going from strength to strength.

Gloucester Old Spot

Looks

The Gloucester Old Spot is a large pig with plenty of meat on it. It has a white coat and large well-defined black spots. Its lop ears drop forward to the nose, but they should not be longer than the nose. The back should be long and level, and the hams should be large and well filled to the hocks. The coat should be silky and straight.

Personality

This is a docile but stubborn pig. Because of its gentle nature, it is an ideal pig for beginners or those with young children.

Habitat

The Gloucester Old Spot should be kept in as free-range an environment as possible. They will benefit from access to grass during summer (see below).

Management

As Gloucester Old Spots tend to put on fat, they do best if allowed to feed on grass alone (with no extra feed) during the summer months. This breed produces meat for all purposes, including a range of pork cuts as well as sausages and bacon.

Breeding

The females make excellent mothers and on average farrow 9–10 piglets. Gloucester sows carry on producing young long after other breeds have stopped.

Gloucester Old Spot sows have a strong mothering instinct.

British Saddleback

docile • pork and bacon pig • good forager • ideal for beginners

TYPE: hardy, traditional breed that is suitable for the production of sucklers, porkers and baconers • SIZE: large • COLOUR: black, with a continuous belt of white hair encircling the shoulders and forelegs

In 1918 herd books for two similar types of pig – the Essex from East Anglia and the Wessex from the New Forest in the south of England – were started. The Wessex is believed to have been a cross of two indigenous English bacon pigs. It was from stock of Wessex pigs exported to the US in around 1825 that the Hampshire (see pages 168–169) was developed. By 1914, the breed was found across south and south-west England, and in the 1940s it was the second most popular breed in the UK.

Both the Essex and the Wessex were predominantly black pigs, but the Essex had white feet and a white tip to its tail, while the Wessex had simply a white band that extended over its front legs. The Essex was believed to be a fancier pig than the Wessex, being known as the 'Gents' Pig', while the Wessex was known as the 'Farmers' Pig'.

In 1967 the two breeds were amalgamated, a new herd book was established and the breed was named the British Saddleback. By this time, like other traditional breeds, Saddleback numbers were declining as more breeders opted for the white breeds.

These days, the breed is enjoying a resurgence in popularity as its

British Saddleback

Its distinctive black-and-white markings make the British Saddleback a very attractive pig.

attractive appearance and docile temperament make it a popular choice for smallholders. Many Saddlebacks are also exported from the UK to countries such as Nigeria, because they perform well in hot climates.

Looks

The British Saddleback is a solid animal with lop ears that should be carried forward without obscuring the animal's vision. Its neck should be of medium length, with shoulders that are of medium width and free of coarseness. It should have a long, straight back with broad and well-filled hams. The coat should be silky and straight. A Saddleback should possess a white band that encircles the shoulders and forelegs – it is deemed a fault for showing purposes if this is not the case.

This grazing pig roots less than other breeds.

Personality

The British Saddleback is friendly, docile and easy-going breed, which makes it an ideal starter pig for novices and a good choice for families.

Habitat

The preferred habitat of this breed is outdoor and free-range. Its dark skin makes it suitable for a range of climates.

Management

The British Saddleback is a hardy breed that adapts easily to all management systems. It grazes well and does not root as much as some of the other breeds. It is an excellent food converter with a high rate of growth, and is capable of reaching mature weights quickly and cheaply. The meat is of very good quality and full-flavoured.

Breeding

The Saddleback is an exceptional mother. This breed is renowned for having plenty of milk and for its excellent mothering ability and for the large, healthy litters it produces, often numbering up to 12 piglets. The sows can be very protective of their litters.

You should aim for a clean continuous saddle in British Saddleback breeding stock.

Berkshire

docile • succulent meat • sunburn-resistant • good mother

TYPE: traditional breed, with a fine dished face and prick ears • SIZE: small • COLOUR: black, with white only on the face, feet and tip of the tail

The earliest surviving record of a Berkshire pig was made in the 17th century by troops under the command of Oliver Cromwell, when they were stationed at Reading in the county of Berkshire. The records speak of good-quality bacon and ham that had come from a locally bred pig.

The original Berkshires were larger than today's more compact breed, and of a black to reddish-brown colour with black or brown spots. In contrast the modern breed is black with white markings on its feet, face and tail.

It has been suggested that the old Berkshires were crossed with Neapolitan pigs to produce the ancestors of today's breed. Chinese and Siamese bloodlines have also been influential crosses, as is shown by the prick ears of the modern breed.

During the 1820s and 1830s, the breed was improved with the help of Lord Barrington. It became so popular in the 19th century that it enjoyed the patronage of Queen Victoria. In 1877 the Royal Smithfield Show at London held separate Berkshire classes, reflecting the popularity of the breed. During the next 13 years, the breed produced 12 Smithfield champions, some owned by members of the royal family.

Berkshires were exported to the United States in the early 19th

Berkshire

The Berkshire's small size and friendly nature have gained it the nickname of the 'Ladies' Pig'.

century, with the first consignments sent to New Jersey in 1823 and New York in 1832, after which the breed spread quickly. Later in the century Berkshires reached Australia and New Zealand, and between the two world wars many pure-breds were exported to Japan.

After the Second World War, commercial breeding companies focused on the white pig breeds and the Berkshire declined in popularity. Only a few loyal breeders kept the pig bloodlines, but today many small-scale pig-keepers once again appreciate the Berkshire's compact size, good looks and docile temperament, and numbers are on the rise.

The story was similar in the United States. The American Berkshire Association kept the breed going through the difficult years and now it is increasingly valued by smallholders and gourmet diners. The breed is still very popular in Japan where its meat is known as the premium 'black pork'.

Looks

The Berkshire's head should be fine with a dished face (upturned nose). Its ears should be largish, carried erect or inclined slightly forward. Its shoulders should be fine and sloping, and its legs should be straight and strong, set wide apart and standing well on its toes. The flesh should be firm, without excessive fat, and the skin should be firm and free from wrinkles. Its hair should be fine and plentiful.

Personality

Known as the 'Ladies' Pig', the Berkshire is placid, friendly and reasonably small. It is an ideal starter breed for novice pig owners.

Habitat

As with all traditional breeds, the Berkshire favours woodland habitats and pasture. This hardy, dark-skinned breed is well suited for keeping outdoors.

Management

This breed is known more for producing pork than bacon.

Breeding

On average, gilts farrow 8–12 piglets while sows farrow 10–12. Berkshires make good mothers and have plenty of milk.

Large Black

hardy • sunburn-resistant • good natured • excellent mother

TYPE: traditional breed with lop ears that is slow to mature, but much appreciated for its succulent taste • SIZE: large • COLOUR: black

The only all-black pig in Britain, the Large Black's hardiness and dark skin make it suitable for export to a range of climates, and by 1935 more than 30 different countries had imported the breed.

The Old English pig is credited with originating the Large Black, which dates back to the 16th and 17th centuries. One theory suggests that pigs mated with Chinese pigs from ships that landed in southwest England to create the forerunners of the breed. Two distinct types developed: the Improved Essex, which was on show in 1840 at the Cambridge Royal Show and was a small fat pig, and the black pigs of Cornwall and Devon, which were larger. In some parts of the world, the Large Black is still known as the Cornish Black.

The 1889 founding of the Large Black Pig Society brought breeders of the two types together and eventually they were amalgamated. The breed's popularity grew and by the early 20th century it had spread through Britain and was being bred with the Large White

Large Black

The breed standard dictates that the ears of the Large Black must reach the tip of its nose.

The Large Black is a hardy outdoor pig that is easy to look after.

(see pages 118–119) and Middle White (see pages 144–145) to produce bacon and pork pigs.

In 1915, the book *Live Stock of the Farm* said the breed's show-ring success was caused 'as much by their meritorious properties as by their great size and weight'. At the 1919 Royal Show, 121 Large Blacks were exhibited, outnumbering other breeds.

The rise in demand for white pigs led to a huge decline in numbers of Large Black in the 1960s. When the Rare Breeds Survival Trust was founded in 1973 the breed was placed on the critical list. Enthusiasts kept the breed going and now it is increasing slowly in popularity. These pigs are also critically endangered in the United States.

Looks

The Large Black is a big pig with long, thin ears that are well inclined over its face, which is why it is sometimes called the 'Elephant Pig'. Its chest should be wide and deep and its back very long and strong. Its hams should be very broad and full, and its tail should be thickset and set moderately high. The coat of the Large Black should be fine and soft with a moderate quantity of black silky hair.

Personality

This breed's superb temperament makes it ideal for novice keepers

Long ears that fall over its face are the cause of the Large Black's nickname: 'Elephant Pig'.

and it is a low-maintenance pig. Like other lop-eared pigs, its view is somewhat restricted, which makes it easier to handle.

Habitat

The Large Black's preferred habitat is outdoors, either pasture or woodland. This is a hardy pig that is suitable for any climate.

Management

This breed is very efficient at converting grass to protein. It grazes and forages well, gaining about 50 per cent of its food intake from the field. Large Black pigs can easily be contained behind a couple of strands of electric fencing. Its meat is succulent and of excellent quality.

Breeding

With their outstanding ability to provide quantities of nutritious milk, the Large Black sow is well known as an excellent mother. Litter sizes of around ten piglets from basic rations are the average. *Guinness World Records* lists a Large Black belonging to A.M. Harris of Lapworth in Warwickshire as having produced 26 litters between 1940 and 1952. The boars are quiet and easy to handle, and they are used as crossing sires over white pigs to give hybrid vigour to porkers or to produce sows for outdoor breeding systems.

Tamworth

mischievous • rooter • talkative • sunburn-resistant

TYPE: alert, traditional breed with a long snout • SIZE: large • COLOUR: ginger

Tamworth

The Tamworth is considered to be the breed most representative of the original indigenous pig that would have inhabited the forests of ancient Britain. Unlike other native breeds, it has not been 'improved' by crossing it with Chinese and Neapolitan stock and it is therefore the oldest pure British breed. It has been crossed with the wild boar (see pages 162–165) to produce the Iron Age (see pages 142–143), for a lean, gamey-tasting pork.

The breed originated in the English Midlands, taking its name from the town of Tamworth in Staffordshire. Due to its ability to produce white-fleshed carcasses with long sides and big hams, the Tamworth enjoyed popularity amongst landowners, farmers, hotel-keepers and cottagers more than a century ago, enabling them to cure their own bacon.

In 1885 the first herd book was started in England. By this time the breed was well established, having received recognition by the Royal Show in England a few years earlier. But after the Second World War breeding stock numbers fell dramatically and during the 1970s

The Tamworth is a strikingly coloured pig with a mischievous temperament.

there were only 17 surviving boars. Britain now looked to Australia to provide new blood, although ironically Australia had initially imported Tamworths from the UK.

Today the Tamworth is going from strength to strength and is growing in popularity as more people hear of its ability to produce excellent bacon.

The Tamworth is a breed known for its love of rooting.

Looks

A Tamworth's head should not be too long. Its face should be slightly dished and wide between the ears, with a light jowl. Its large ears are finely fringed and carried slightly inclined. Its skin should be flesh-coloured and free from coarseness, wrinkles or black spots. The coat should be a distinctive golden red, abundant, straight, fine and as free as possible of black hairs.

Personality

This mischievous pig is full of character and a good escape artist. It is not for the beginner.

Habitat

The Tamworth was originally a woodland pig, but will do just as well on pasture. It should be kept as free-range as possible. This hardy breed does well in any climate and is resistant to sunburn.

Management

Primarily a bacon pig, the Tamworth can also be used for pork.

Breeding

The sows are good mothers, not aggressive and protective of their litters. They produce plenty of milk. The litters are small, between five and nine piglets.

Tamworth litters contain between five and nine piglets, fewer than those of other breeds.

Iron Age

unusual breed • rooting pig • lean, gamey meat • protective mother

TYPE: hybrid breed that can be quite a handful • SIZE: medium • COLOUR: depending on the percentage of wild boar, it can be reddish like the Tamworth or greyish brown like the wild boar; some are nearly black

The Iron Age pig is a hybrid of the Tamworth (see pages 138–141) and the wild boar (see pages 162–165). It was established in the 1970s as the result of a project to create a pig similar to those that would have lived during the Iron Age. When a Tamworth sow was crossed with a European wild boar from London Zoo the offspring – the Iron Age pig – did indeed look very much like an ancient pig.

These hybrids are much tamer than the wild boar, but not as people-friendly as domestic pigs. Initially, they were only seen in farm parks, but now more smallholders are realizing their benefits. Their meat is much leaner and gamier than that of other pigs, and is often sold through specialist outlets and farmers' markets.

Iron Age

Looks

The Iron Age is an aggressive-looking pig that appears more frightening than it really is. It is medium-sized, with bristly hair, prick ears and a long snout, plus a strikingly deep ribcage.

Personality

Iron Age pigs are not suitable for a beginner. They possess the characteristics of both of their parents, being quite active and good escape artists. When you are in a pen with them, they must be watched carefully.

Habitat

This breed must be given a free-range environment. Their natural habitat is woodland, but the pigs will also do well on pasture.

Management

Iron Age pigs are difficult to keep. They have a different behaviour pattern to traditional breeds, being very destructive, and will root more than any other pig (even Tamworths). Shelters must have a concrete floor, otherwise they will destroy them. Excellent fencing is also a must with this breed.

Breeding

The Iron Age sow tends to have small litters of stripy piglets, which lose their stripes as they mature. The mothers are very protective.

Iron Age piglets have an attractive stripy coat that they lose as they grow older.

Middle White

grazer • docile • delicious meat • good mother

TYPE: traditional breed, smaller and more compact than other breeds, well fleshed and early maturing • SIZE: medium • COLOUR: white

The breed came into being quite suddenly in 1852 when Joseph Tuley, who was a weaver from Keighley and a prominent breeder of Large Whites (see pages 118–119), entered several of his pigs at the Keighley Agricultural Show in Yorkshire. The judges felt that the pigs were neither of sufficient size to be Large Whites nor correct for the Small White, so decided to set up a third class for the 'Middle Breed'. To establish the breed, Tuley crossed his best Large Whites with Small Whites.

By 1912 the Small White had disappeared. This breed, which had been developed for showing from imported Chinese and Siamese pigs, left its legacy on the Middle White in the dished face that is the cause of its nickname, the 'Bat Pig'.

Thanks to its early maturing and ease of management, the Middle White rose in popularity. Along with the Large White and the Tamworth (see pages 138–141), the Middle White was a foundation breed of the National Pig Breeders'

Middle White

Association, which was founded in 1884. The first Middle White herd books were published in the same year.

A favourite with butchers, especially in London, the breed became known as the 'London Porker'. The carcass lent itself to being cut into the small pork joints that were much in demand in the early 1900s. However, along with other specialist pork breeds, the Middle White declined in numbers during the Second World War when rationing created a demand for the bacon pig.

Today the Middle White is popular once again. The UK's Middle White Pig Breeders' Club, which was established in 1990, has as its patron the well-known chef Antony Worrall Thompson, who breeds his own Middle White pigs. Other countries have also found the Middle White to be a valuable breed and it has been exported all over the world. In Japan where it is very popular, it is known as the Middle Yorks.

The Middle White is known as the 'Bat Pig' for its dished face with upturned snout.

Looks
The Middle White is an all-white pig with a thin skin and a fine coat. It has prick ears and a short, dished face with an upturned snout.

Personality
The Middle White is a docile, easily managed pig and makes an ideal beginner's breed.

Habitat
Like other traditional breeds, this pig should be kept in an outdoor environment in pasture or woodland. Herds are sometimes used in forest management.

Management
The Middle White tends to graze rather than root, but because of its thin skin, it must have warm shelter during the winter. It is regarded as a specialist breed for suckling pigs and is much in demand in the restaurant trade.

Breeding
The sow is a good mother with plenty of milk, and usually farrows around nine or ten piglets. Boars are often crossed with other breeds for pork or bacon production.

Welsh

hardy • lean • docile • easy to keep

TYPE: placid, modern breed with a wide back and well-muscled loins • SIZE: medium • COLOUR: yellow-white

There are records of Welsh pigs being traded from the 1870s into Cheshire, where they were fattened on the by-products of the dairy industry and crossed with Manchester boars.

When imports were restricted by the First World War, demand for home-produced pigs rose and this benefited the Welsh breed, among others. The first breed society in Wales, the Old Glamorgan Pig Society, was established in 1918. Pigs similar to those of Glamorgan were also being produced in other parts of Wales, so in 1920 the Welsh Pig Society was formed and in 1922 it was amalgamated with the Old Glamorgan Pig Society. The first herd book followed in the same year.

After the Second World War, the Welsh, like other pigs, increased in number thanks to better supplies of animal feed. In 1952 the Welsh breed joined the six other pedigree breeds already represented by the National Pig Breeders' Association (now known as the British Pig Association). The breed became increasingly popular: for example, pedigree sow registrations increased from 850 in 1952 to 3,736 in 1954. The Welsh became the number three sire breed in Britain, with the Large White (see pages 118–119) at number one and the Landrace (see pages 120–123) in second place.

In the 1980s registrations declined, but as the breed is valued in cross-breeding for characteristics such as its ease of management

Welsh

and hardiness, its popularity is once more on the increase. The breed currently has 14 boar lines and 32 sow lines.

Looks

The head should be light, fine and fairly wide between the ears. The back should be long, strong and level, with well-sprung ribs giving a fairly wide mid-back. The tail should be thick and free from depressions at the root. The loins should be well muscled, firm and well developed. The hindquarters should be strong with hams that are full, firm and thick. The coat should straight and fine.

Personality

The Welsh is a quiet and easily managed pig that is suitable for beginners.

Habitat

This breed works well in both outdoor, free-range environments and in indoor commercial systems. If kept outdoors, this pale-skinned breed needs plenty of shade.

Management

If kept outside in summer, it must be watched closely for sunburn.

Breeding

The sows are natural mothers and usually farrow between one and 12 piglets. Welsh pigs are in demand for cross-breeding programmes.

The Welsh is gaining popularity nowadays outside its native Wales.

British Lop

docile • suitable for pork or bacon • reasonably lean • good grazer

TYPE: long-bodied, traditional breed • SIZE: large • COLOUR: white

The British Lop was first recognized close to the town of Tavistock in the West of England, where it was known as the Devon Lop or Cornish Lop. For most of its history it remained a local breed, popular with farmers in the surrounding area. Even when it spread further into England, it was rarely seen beyond Somerset and Dorset. In those days it was known as the National Long White Lop Eared breed, until its name was changed to British Lop in the 1960s.

In 1973 the British Lop was listed as one of the six rare breeds recognized by the Rare Breeds Survival Trust. Over the last few years, the breed's numbers have grown as more people appreciate the quality of the meat and its good nature and hardiness, although it is still considered endangered.

British Lop

Looks

The British Lop is long and lean, with large hams that are well filled to the hocks. It has long, fine white hairs and lop ears. Its back should be long and level, with a strong tail set high. Its legs should be of medium length, straight and set level with the outside of the body.

Personality

Docile and easily managed, the British Lop is suitable for beginners.

Habitat

This traditional breed prefers an outdoor environment (woodland or pasture), but it will also do well in commercial indoor systems. If kept outdoors, provide shade during summer to protect against sunburn.

Management

This pig is suitable for both small-scale and extensive commercial systems and produces pork and bacon of excellent quality. It grows quickly and will finish with a well-muscled lean carcass.

Breeding

The sows are easy farrowers with plenty of milk. They usually produce 15–16 piglets and rear 12–13. The boars can improve other breeds when used as terminal sires.

A reasonably lean breed, the British Lop is suitable for both pork and bacon.

Pietrain

docile • very lean meat • improver of other breeds • suitable for intensive management

TYPE: lean, highly muscled, modern breed with great hams • SIZE: medium • COLOUR: white with black patches

Pietrain

This breed takes its name from the village of Pietrain in Belgium where it was first developed. In 1920 Gloucester Old Spot boars (see pages 126–127) were sent over to breed with local pigs, in the hope of improving them. In 1947 a Gloucester Old Spot sow called Espérance de la Sarte arrived in the village of Pietrain and she and her progeny were the foundation for the Pietrain herd book that was established in 1953. The village was closed to all outside pigs to ensure the purity of the breed. Nearly all of the Pietrain pigs in the herd book can trace their ancestry to Espérance.

The breed became popular in its native country and subsequently was exported to other European countries, including Germany and Spain, where it was used extensively as a terminal sire. The first Pietrain pigs arrived in the UK in 1964.

The Pietrain is renowned for its very high yield of lean meat and is therefore often used in genetic improvement programmes. In the past it has been associated with the presence of the gene for Porcine Stress Syndrome, but this has now been bred out.

Looks

A medium-sized pig with erect ears, the Pietrain is clearly recognizable by its enlarged, muscular shoulder mass and its fully muscled back.

Personality

This is an extremely docile pig, with both sows and boars being easy to handle.

Habitat

This breed is usually found in commercial indoor systems, but there is no reason why it can't be kept in an outdoor environment.

Management

An excellent improver of other breeds, the Pietrain tends to be used commercially rather than being kept by smallholders. Animals of this breed need to be supplied with high-quality feed.

Breeding

The Pietrain is an excellent mother with plenty of milk and farrows litters that average 12 piglets.

The Pietrain is a lean, very muscled pig that makes a good mother.

Swabian-Hall Swine

docile • superior meat • sunburn-resistant • good mother

TYPE: docile breed with lop ears • SIZE: large • COLOUR: black and white

Swabian-Hall Swine can trace their ancestry back to Meishan pigs (see pages 190–191) crossed with Germany's indigenous breeds. In 1830, King Wilhelm I imported Chinese Meishan pigs to cross with local pigs, in order to increase the fat content of the resulting progeny. The breed was embraced by local farmers, who appreciated the superior quality of the meat. Darker than that of other pigs, the pork has a strong flavour and distinctive smell.

By the mid 1840s the breed was also very popular in other German-speaking countries. It remained successful in Germany for a further century: in the late 1950s, 90 per cent of pigs in Baden-Württemberg were Swabian-Hall. However, by the end of the 1960s, due to the fat content of the meat, its popularity was declining fast. After the breed society closed, it was only a few determined breeders who kept the Swabian-Hall Swine going.

In 1980 the breed found a new lease of life and a new breed society was started in 1983. Since then, although still endangered, the Swabian-Hall has gone from strength to strength. Its succulent meat is today much in demand by good restaurants.

Swabian-Hall Swine

Looks

The Swabian-Hall is a large pig, predominantly white with a black head and rump. It also has zones of grey colouring, due to the effect of white hairs seen against black skin pigmentation.

Personality

This docile pig is easily managed.

Habitat

It is recommended that this pig is kept on pasture during the summer

Only around 900 Swabian-Hall Swine exist and it remains an endangered breed.

months, although it can be brought inside during winter. A hardy pig, it is suitable for a range of climates.

Management

The Swabian-Hall is found on farms affiliated with the Bäuerliche Erzeugergemeinschaft Schwäbisch-Hällisches Schwein (Farmer Producer Association Swabian-Hall). Member farms are regularly inspected to ensure high standards of welfare and feeding based on GM-free cereals, peas and bean meal. Owners are required to keep pigs at pasture during summer, and the animals may not be given any drugs or antibiotics.

Breeding

The sow normally farrows around nine or ten piglets. The Swabian-Hall is a good mother with a plentiful supply of milk.

Iberian

docile • good forager • suitable for curing • excellent flavour and marbling

TYPE: dark, almost hairless breed that forages on acorns • SIZE: large • COLOUR: black or red

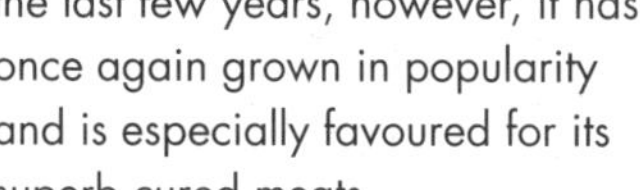

It is thought that the original Iberian pigs were brought to the Iberian peninsula from the eastern Mediterranean coast. There, they interbred with wild boars and this cross gave rise to the breed we know today, whose origins can be traced back to 1000 BC.

The Iberian has declined in number since the 1960s due to an outbreak of African swine fever, the introduction of leaner and faster-growing foreign breeds and a reduced market for animal fats. In the last few years, however, it has once again grown in popularity and is especially favoured for its superb cured meats.

The Iberian has the ability to accumulate fat under the skin as well as between its muscular fibres. It is this fat that produces the white streaks, or marbling, that gives the ham its special taste.

Looks

The Iberian pig is dark in colour and nearly hairless.

Iberian

Personality

This is a placid pig that can live in groups without fighting.

Habitat

The ideal habitat is woodland and pasture. Traditionally, holm or cork oaks would be present (see below). Its dark skin makes this breed suitable for a range of climates.

Management

Iberian pigs consume a diet of acorns from the holm and cork oaks. They will normally consume 6–10 kg (13–22 lb) of acorns a day, as well as about 3 kg (6½ lb) of grass. This diet is essential if you are keeping the breed for its meat, as the quality of the hams depends on the acorns fed to the pigs. During weaning the Iberian is fed on high-quality fodder to enable it to reach an ideal weight. The breed is much sought-after for its carcass joint, hams, forelegs and loins.

Breeding

Iberian pigs usually have only around six piglets a litter and farrow twice a year, in March and October.

The Iberian is famous for the quality of meat it produces, particularly its Parma ham.

Mangalitza

docile • carries a high degree of fat • good forager • three colours available

TYPE: large-framed, dark-skinned, curly-coated breed • SIZE: medium • COLOUR: swallow-bellied (black with a white underbelly), blonde (grey to yellow) or red

Mangalitzas, otherwise known as Woolly Pigs, are found in Austria, Germany, Hungary, Romania and Switzerland. During the 1900s, a large number of Lincolnshire Curly Coats were exported to Hungary to improve the local Mangalitza breed. The Lincolnshire Curly Coat became extinct in the early 1970s but still lives on in the Mangalitza.

Mangalitza meat proved to be so popular in Europe that it was once traded on the Vienna Stock Exchange. However, later in the 20th century the breed nearly became extinct, with the worldwide population down to fewer than 150 sows in 1993.

Once known for its qualities as a lard pig producing 70 litres (123 pints) of rendered fat, today's Mangalitza is made into special meats and salamis, as well as being used in forestry projects. In 2004 the breed was featured at the Salone Del Gusto in Turin and it is now starting to appear regularly at shows in the UK.

Male and female Mangalitzas of the three colours were imported

Mangalitza

The blonde Mangalitza is very similar to the now-extinct Lincolnshire Curly Coat.

into the UK in 2006. Seven female lines and three boar lines were established and the number of Mangalitza are slowly growing as more and more people are buying piglets and breeding them. All three types of Mangalitza are very hardy and docile. Their curly coat enables them to stay out in the hot sun without getting sunburn.

The swallow-bellied Mangalitza was developed in the 1800s by crossing the blonde Mangalitza with the black Mangalitza. However, the black pig became extinct in the 1970s, with the last-known herd being found on Serb islands in the Danube.

The red Mangalitza is a slightly larger pig in size and weight, with a thick reddish coat.

Looks

The Mangalitza's fur is dense and long, curling in winter and becoming shorter and straighter in the summer with seasonal moulting. The head is of medium size, with ears that lean forwards. The underline should have a minimum of ten teats, five on either side. The back of this breed should be straight or slightly arched.

The Mangalitza's curly coat makes it suitable for a range of climates.

Personality

Both Mangalitza boars and sows are extremely docile, friendly pigs, which like being around people. They are suitable for beginners.

Habitat

They are very hardy outdoor pigs and their ideal environment is free-range woodland or pasture. The curly coat of the Mangalitza allows it to live comfortably in the snows of Austria, while their pigmented skin also makes them suitable for hotter climates.

Management

Compared to the more modern commercial pigs, the Mangalitza is slow growing and can take up to two years to reach its ideal weight of 100 kg (220 lb). The slow maturing of this breed means that the meat is of exceptional quality and greatly in demand for the production of air-dried and cured meat. Their wool is also used for making flies for fishermen. The breed is suitable for smallholders.

Breeding

Mangalitzas are good mothers that produce small litters of around six or seven piglets, which are striped in a similar way to the boarlets of the wild boar (see page 164).

This attractive breed is available in three different colours.

Bentheim Black Pied

hardy • endangered • meat and lard pig • good mother

TYPE: hardy, docile breed that is easy to manage • SIZE: medium • COLOUR: white with black spots in grey rings

Also known as the Spotted German Pig, the Bentheim Black Pied originated in Bentheim in northern Germany after being crossed with local pigs and with the Berkshire (see pages 132–133).

The Bentheim Black Pied breed is endangered. At one point, for a period of 25 years, only one farmer was breeding them, and the herd book was closed in 1964. It was re-established in 1987 and a breed society was founded in 2004–2005 to try to augment numbers. Today, there has been an increase to a population of 380 Bentheim Black Pieds.

Looks

The Bentheim Black Pied is medium-sized and white in colour, bearing black splodges with grey rings.

Personality

This is a docile breed.

Habitat

The natural habitat of this pig is the lowlands of northern Germany. It is a hardy breed and can comfortably live outdoors, on either woodland or pasture.

Management

Bentheim Black Pieds are usually kept on small farms with two or three breeding sows.

Breeding

The sows are good mothers and can farrow up to 14 piglets.

Bentheim Black Pied

A Bentheim Black Pied sow will produce as many as 14 piglets in a litter.

Wild boar

shy • slow to mature • lean meat • semi-domesticated

TYPE: long-snouted pig with a line of upright hairs along its back • SIZE: medium • COLOUR: greyish black

The wild boar was originally found in most areas of the world, but today has greatly decreased in numbers, although it is still common in Europe (especially in the forests of Germany, Denmark and France), North Africa and Asia.

It is thought that the wild boar became extinct in Scotland during the 17th or 18th century because it was hunted for sport and meat, and because the forests that had provided its home were cleared. In England the wild boar is believed to have died out earlier than that, although unsuccessful attempts were made to reintroduce it in the 17th century.

When their enclosures were damaged during storms in 1987 and 1989, wild boars escaped from boar farms and started to breed. Today there are small colonies dotted all over the UK, although no one really knows for sure quite how many boars are living wild. Every year in Germany

Wild boar

Semi-domesticated animals, wild boar can be aggressive and are difficult to keep.

a million wild boars are shot for sport or as pests.

The wild boar has a good turn of speed and impressive stamina. It was regularly hunted, especially by royalty, and it is believed that a wild boar nearly gored Henry VIII to death during one such hunt. In her 1861 *Book of Household Management*, Mrs Beeton wrote: 'The Boar's head, in ancient times, formed the most important dish on the table, and was invariably the first placed on the board at festive feasts, being preceded by a body of servitors, a flourish of trumpets, and other marks of distinction and reverence, and carried into the hall by the individual of next rank to the lord of the feast.'

The wild boar has acute hearing and sense of smell, although its eyesight is poor. Normally quite a shy animal, it will only become aggressive if it is guarding its young – fighting to the death if necessary – or if it is cornered; otherwise it prefers to stay well hidden and away from humans.

Looks

The wild boar has a long body and short legs. It has short, bristly fur, a prominent snout and a ridge of longer hair along its back. The tail is short and tasselled.

Personality

Wild boar retain more of a natural instinct than other pigs and can be aggressive. They are definitely not pigs for the novice. In the wild, males are solitary creatures and will often hunt on their own, but wild boar females are sociable and live in groups known as soundings. They are highly vocal and talk to one another in grunts and squeals.

Habitat

Wild boar are incredibly hardy and can live outdoors all year round. Owners should provide a woodland environment as far as possible, rather than pasture, as this replicates the wild boar's natural surroundings.

Management

Wild boar are kept solely for their meat. They grow slowly, taking between 12 and 14 months to mature, and because of this slow growth they are quite costly to farm. They live as a family unit, with the older sows ruling the unit. In the UK, a wild-animal licence is needed to keep them.

Breeding

Female wild boar live as a herd, with some sows even babysitting for piglets other than their own. The piglets stay with the sow until she gives birth again, even when they have been weaned. Females usually give birth to between four and seven piglets in each litter. At birth the young (known as boarlets) have stripes, which they later lose.

Wild boar need to be allowed access to a free-range outdoor environment.

American Yorkshire

docile • muscular carcass • excellent feed convertor • good mother

TYPE: muscular breed with a high proportion of lean meat • SIZE: large • COLOUR: white

The American Yorkshire was originally developed in the English county of Yorkshire during the 18th century. In Britain this breed was later to become known as the Large White (see pages 118–119). In around 1830, Yorkshire pigs were first imported to Ohio in the United States. However, due to their initially slow rate of growth, they did not at first find favour with American farmers, despite heavy promotion by the Morrell Packing Company of Ottumwa, Iowa, and the Hormel Packing Company of Austin, Minnesota.

The American Yorkshire breed was initially more popular in Canada, where the pigs were in demand for their hardiness and the muscular quality of the carcass. However, farmers in the United States gradually began to realize the full potential of the Yorkshire and started to use the pigs as breeding stock. In the 1940s the breed expanded rapidly when a large number of Yorkshires were imported to the United States from Canada and England.

American Yorkshire

The Yorkshire is one of the most popular breeds in the United States and Canada.

Today pure Yorkshire pigs are found in nearly every American state, as well as appearing in crosses all over the world.

Looks

American Yorkshire pigs are white, with erect ears. Muscular and well fleshed, they have a high proportion of lean meat and low back fat.

Personality

This is a docile breed.

Habitat

The Yorkshire is usually found in commercial indoor systems.

Management

Selection and imported stock have created a lean, fast-maturing breed that meets the needs of the modern pork producer. In summer pigs that are kept outdoors should be watched for sunburn.

Breeding

Yorkshires have a natural mothering ability, and will happily raise 12–14 piglets in a litter. They are caring mothers with plenty of milk.

Hampshire

easy to handle • prick ears • good forager • lean meat

TYPE: strong-boned, modern breed • SIZE: large • COLOUR: predominantly black, with a white saddle and front legs

The *Hampshire Blue Book*, which was published in 1928, describes the history of the breed, including how it was exported from the English county of Hampshire to the United States in 1832. In the United States the breed became known by a range of different names, including McGee Hog, Saddleback, McKay and Ring Middle. In 1890 it was decided to use the name that represented the breed's origins – the Hampshire.

The Hampshire was also given the nickname of 'Thin Rind', due to its famously muscled carcass which contrasted with the fat carcasses of the lard-producing pigs that were more common at that time. Because it is the leanest of the North American breeds, the Hampshire is used in many countries as the sire of pigs cross-bred for pork. In the United States, many of the carcass competitions are won by Hampshires and Hampshire crosses.

In 1968 the Hampshire was imported from the United States into the UK by the Animal Breeding Research Association. Forty more pigs, this time from Canada, joined these in 1973. The breed quickly

Hampshire

became popular worldwide and 600 British Hampshires were exported to 14 different countries from August 1978 to August 1979.

At the Royal Smithfield Show in London, the lean British Hampshire and its crosses has won many interbreed championships, both for live animals and carcasses. The breed is regarded as one of the best terminal sires in the commercial world.

Looks
This breed has a medium-sized head with prick ears and a long snout. It should have long, deep sides with well-sprung ribs, long legs and a tail set high.

Personality
The Hampshire is easy to handle, despite its size.

Habitat
The Hampshire is suited to both commercial indoor systems and free-range outdoor environments.

Management
This hardy pig does well in any environment and is a good forager.

Breeding
Hampshires make excellent mothers with plenty of milk, farrowing large, healthy litters of 8–10 piglets.

The Hampshire is similar in appearance to the British Saddleback.

Duroc

hardy • sunburn-resistant • matures quickly • protective mother

TYPE: extremely hardy modern breed that is usually seen in commercial enterprises and is often used as a terminal sire or in hybrids • SIZE: medium • COLOUR: different shades of red

Duroc

The Duroc probably originated from importations into the United States at the start of the 19th century of Red Guinea pigs from Africa and Spanish Red and Red Portuguese pigs from Europe. Red pigs were being bred in New York and New Jersey by 1812 and were known for being large animals that produced good numbers of piglets.

The Duroc of today is a mixture of the Red Duroc from New York and the New Jersey version, the Jersey Red. The Red Duroc was named after a famous stallion called Duroc. In 1823 Isaac Frink of Saratoga County, New York, bought piglets and a red boar from Harry Kelsey, owner of the stallion. The pigs were named Duroc in the horse's honour and the resulting progeny inherited the name along with the red colouring, pleasant nature and good growth rates.

The Jersey Red strain was developed by Clark Pettit. The name was given to the breed by the agricultural editor of the *New York Tribune*, who lived in New Jersey. The Jersey Red was a coarser

The hardy Duroc can adapt to any environment, indoor or outdoor.

This active pig enjoys the company of humans.

and longer pig than the Red Duroc, and the latter was known for having a better quality of carcass.

In the 1880s the Duroc breed became well established throughout the corn belt. The breed society, the American Duroc-Jersey Association, was formed in 1883 with the aim of improving, promoting and recording red pigs. In 1893 the first Duroc Hog Show was held at the Chicago World's Fair, helping to increase the breed's popularity worldwide. In 1934 the groups amalgamated to establish the United Duroc Swine Registry.

The first attempt in the early 1970s to import Durocs into the UK was not very successful, although some pigs were later exported to Denmark. In the 1980s, further imports, backed up by trials to see how the pigs performed as a terminal sire, were more successful. Further research also established the tenderness of the Duroc's meat.

The Duroc's thick auburn winter coat turned out to be ideal for the emerging outdoor pig market, helping the animals to cope with the British winter. In summer the pig moults, allowing it to deal more effectively with heat. The Duroc therefore has a special place in British pig-keeping and a unique British version has developed.

Looks
The Duroc's head is small in proportion to the size of its body. Its ears should be of medium size and moderately thin, pointing forward, downward and curved slightly outward. Its shoulders should be quite broad and very deep and full. Its back should be of medium breadth, straight or slightly arching, and its hams should be broad and full.

Personality
An active pig that is full of character, the Duroc is easy to maintain and very hardy. Its docile nature makes it ideal for beginners.

Habitat
The Duroc may be kept either outdoors or in a commercial indoor system. Exceptionally hardy, it can adapt to any environment.

Management
The succulence and heavy muscling of the Duroc make it suitable for anything from light pork to heavy hog production. This is the second most-recorded breed of swine in the United States, popular for its meat and ease of management.

Breeding
The Duroc has plenty of milk and is an excellent mother that farrows large litters of 8–12 piglets. The sow can be very protective of her piglets.

The Duroc's red colouring means it is resistant to sunburn.

Lacombe

docile • meaty • high fertility • large litters

TYPE: modern breed that is docile and easy to manage • SIZE: large • COLOUR: white

The breed is named after the Lacombe research station in Alberta, Canada, where the pigs were developed. In 1947 breeding began at this Department of Agriculture research station to develop a cross using the Chester White (see pages 178–179), the Danish Landrace (see pages 120–123) and the Berkshire (see pages 132–133). The project took 11 years to complete. The new breed was intended to replace the Landrace, and the foundation breeds used were chosen for their desirable characteristics: docility, rapid weight gain, large litter sizes, efficiency of food conversion and carcass quality.

The foundation breeds were crossed and back-crossed to form a pure composite breed, with selection based on litter size, quality of the carcass and weight gain. Any pigs that were found to have coloured skin, or that did not measure up to the rigorous criteria, were culled. This resulted in a pure-breed Lacombe with 56 per cent Landrace, 21 per cent Chester White and 23 per cent Berkshire genes.

A public exhibition of the newly developed Lacombe was held at the National Swine Show at Brandon on 1 July 1957. In October of the same year, the breed was released to the public after positive results following rigorous testing of the Lacombe's performance in cross-breeding and in pure-bred experimental herds. A national draw was organized for the first 50 boars, and a few years later the first Lacombe sows were made available to the public.

In 1958 the Lacombe was first registered with the Canadian National Livestock Records, and by 1960 it was established as a top-rank pig; by 1969 it had been imported into more than eight other countries. Today, however, the number of breeders owning Lacombes can be counted on one hand, as the focus is on cross-breds rather than pure-breds.

Looks

The Lacombe looks similar to a Landrace, but has a slightly heavier bone structure and lop ears. The breed is medium-sized and long-bodied, with short legs.

Personality

The Lacombe is an extremely docile breed and it is suitable for beginner pig-keepers.

Habitat

The Lacombe will thrive in both indoor commercial systems and outdoor environments such as woodland or pasture. Like other white-skinned pigs, Lacombes could be susceptible to sunburn so need protection in hot climates.

Management

The Lacombe is currently farmed by a small number of independent breeders dedicated to ensuring, through selective breeding, that the breed continues to deliver the desirable characteristics demanded by the market. For example, the sows are of extremely docile temperament and produce litters that are easy to manage in an intensive environment and gain weight rapidly.

Breeding

The Lacombe is a good mother with plenty of milk, with an average of 9–12 piglets.

Lacombe

Poland China

hardy • active • good weight gain • excellent feeder

TYPE: big-framed, long-bodied, lean, muscular breed • SIZE: large • COLOUR: black with a white snout and four white stockings

It is believed that the Poland China derives from stock brought into the Miami valley of the United States by early settlers. It is uncertain exactly where this stock originated. There is evidence that some of these early pigs were Berkshires (see pages 132–133), while some of the stock came from Kentucky and was probably of the same breeding as those pigs that later became known as the Hampshire (see pages 168–169).

In the past, very large sows were being used for breeding Poland Chinas, and as a result extremely large litters were produced, sometimes with 16 or 17 piglets. However, the sows were so large that they made poor mothers, squashing many piglets in the first days of farrowing. Since then, the size and mothering abilities of the sows have been improved by means of selection.

Looks

Over the years the Poland China has been developed from a fatty pig to a leaner breed. It is very similar in colour to the Berkshire, but has drooping ears and a flatter, longer topline. Some Poland Chinas have white spots on different parts of the body, particularly on the nose, feet and tail.

Poland China

Personality

This pig is active yet docile, and therefore suitable for beginners.

Habitat

This hardy breed enjoys an outdoor habitat of woodland or pasture and is not suitable for commercial intensive systems.

Management

The breed is known for its natural soundness, substantial bone and the lean carcass that makes it an ideal pork breed. Sows are often used to improve other breeds in cross-breeding programmes.

Breeding

Poland Chinas make reasonable mothers with plenty of milk, and they stay in good condition throughout lactation. The average size of their litters is nine piglets.

Spots

active • matures early • prolific breeder • good weight gain

TYPE: quick-growing breed with feed-efficient qualities • SIZE: large • COLOUR: large black and white spots over its body

For many years this breed was known as the Spotted Poland China. The three foundation stocks were Gloucester Old Spot (see pages 126–127), Poland China (see page 176) and the spotted pigs of local origin.

The breed originated in Putnam and Hendricks counties in Indiana, where three breeders occasionally brought sows and boars from Ohio, which they then crossed with their own pigs, in time developing their own breed. In 1960 the name of the National Spotted Poland China Swine Record was changed to the National Spotted Swine Record by members of the association. It was also agreed that the name of the breed would change to Spotted Swine (or Spots for short).

Spots are popular with both farmers and commercial producers for their ability to produce piglets with a fast growth rate.

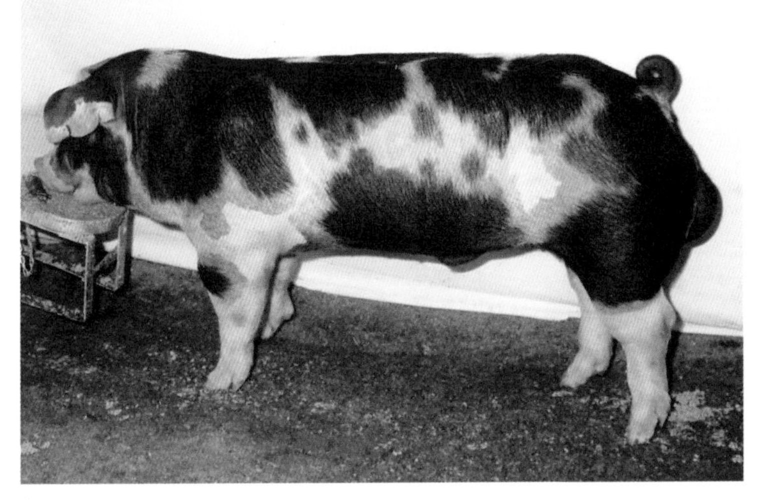

Spots

Looks

The Spots is a stocky, muscular pig. It should have a level topline, a medium-length snout and ears that fall forward.

Personality

Although generally good-natured, this active breed can be aggressive with other breeds of pigs. These are not considered pigs for novices.

Habitat

The Spots is suitable for both commercial indoor systems and outdoor environments, such as pasture and woodland. It is suitable for a range of climates.

Management

This breed is becoming increasingly popular with commercial producers as the quality of its carcass improves. They need to be kept away from other breeds of pigs.

Breeding

Spots are protective mothers, raising 9–10 piglets in a litter. The piglets usually show a good daily rate gain.

Chester White

docile • large litters • slow-growing • good mother

TYPE: hardy white breed used for meat • SIZE: large • COLOUR: white

Between 1815 and 1818 a white boar, possibly a Bedfordshire or Cumberland, was exported from Britain to the US and used to develop a white breed of pig. The Chester White was originally called the Chester County White after the place where it originated – Chester County, Pennsylvania – and the word county was later dropped.

Thomas Sharpless established the Chester White Record Association in 1884, and in the years that followed other Chester White Associations and recording offices developed across the United States. Around 1911, F.F. Moore of Rochester, Indiana, organized the amalgamation of the various associations into the Chester White Swine Record Association, which was based in Lima, Ohio.

F.F. Moore became the breed association's first president. In 1910 his three sons started the first breed magazine, the *White Breeders' Companion*, which changed its name to the *Chester White Journal* in July 1918.

More than 60,000 Chester Whites are recorded in the United States each year, and this is probably not a true representation of the total number of Chesters because many are never recorded.

Chester White

The Chester White is renowned for the tenderness and succulent taste of its meat.

Looks
The Chester White is a medium-sized pig with semi-lop ears and a white coat and white skin.

Personality
The Chester White is a docile pig and therefore suitable for novice pig-keepers.

Habitat
The preferred habitat of the Chester White is outdoors in woodland or pasture, but it may be kept in commercial indoor systems.

Management
Chester Whites are highly valued for their hardiness and the marbling and tenderness of their meat (which is also known for its easy slicing).

Breeding
Producing litters of between 12 and 14 piglets, this breed is a world leader in numbers of piglets per sow per year. The sows are known for their natural mothering ability.

Hereford

docile • matures early • good beginner's pig • prolific breeder

TYPE: gentle breed that is ideal for small-scale farming • SIZE: medium to large • COLOUR: white face with at least two white feet and a red body

The Hereford was first developed in Iowa and Nebraska during the 1920s, from the Duroc (see pages 170–173), the Chester White (see pages 178– 179) and the Poland China (see page 176). In 1934, approximately 100 animals were selected as foundation stock and the National Hereford Hog Record was formed to promote this new breed. Within the first ten years the new breed association attracted 450 members, as the Hereford breed became popular with the local cattlemen.

During the 1960s, however, the Hereford began to decline, due to increasing interest in the crossing of the Duroc, Hampshire and Yorkshire breeds. Today fewer than 2,000 Hereford remain. Most breeders are in the Upper Midwestern states.

Hereford

Looks
A medium- to large-sized pig, the Hereford's face is of medium length, with a slight dish (upturned nose) and medium-sized ears. Its body should be red – the darker it is, the better.

Personality
The Hereford is a very docile breed and it is therefore ideal for the novice pig-keeper.

Habitat
This breed does best on pasture or on small-scale farms. It will do well indoors as well as outside, but is not ideally suited for intensive commercial systems. The Hereford will thrive in most climates.

Management
The Hereford is an early maturer and is very efficient at converting food into growth. It also crosses well with other breeds.

Breeding
Herefords make excellent mothers and produce and wean large litters of around 10–14 piglets.

The attractive and easy-to-manage Hereford will do well on pasture.

Red Wattle

active • tasselled pig • meat darker than other pork • large litters

TYPE: large breed with unique tassels, known for its lean pork • SIZE: large • COLOUR: different shades of red, sometimes with black specks or patches; some pigs are nearly black

The Red Wattle is so called because of its ruddy hair and fleshy wattles or tassels, which are the result of a single gene. It is believed that the breed was originally brought to the United States from New Caledonia, a French island in the South Pacific. Up until the 1700s the breed could be seen in French New Orleans. Today's Red Wattles are thought to be derived from wild Red Wattle pigs that breeders discovered living in wooded areas in eastern Texas in the early 1970s.

In the 1980s both breeding and market pigs fetched a premium, and pigs crossed with Red Wattles brought about a leaner hybrid. Three organizations kept registries of the Red Wattle, but there was never an active breed association. The American Livestock Breeds Conservancy tried to encourage breeders to combine their efforts for the good of the breed, but the organizations remained separate and for many years no new stock was registered. In 1999 it was discovered that only 42 breeding animals remained, belonging to six breeders.

Like so many traditional breeds, the Red Wattle had fallen out of favour with pork producers due to the rise in popularity of the more commercial breeds and the lack of customers asking for its meat.

Red Wattle

Looks

The Red Wattle is a large pig with a fleshy wattle attached to each side of the neck. It should have a lean head and slim snout. Its ears are upright with a drooping tip. The body is compact and the back is slightly arched. It measures about 2.4 metres (8 feet) long and about 1.2 metres (4 feet) high.

The fleshy tassels or wattle on either side of the Red Wattle's neck make this a distinctive pig.

Personality

This active pig can be a handful, so is not for the absolute novice.

Habitat

The preferred habitat of the Red Wattle is outdoors in pasture or woodland. They are suitable for a wide range of climates.

Management

A good forager and very hardy, the Red Wattle is ideal for small-scale production. Its colouring protects it from sunburn. The meat is very lean and juicy, with excellent hams.

Breeding

The sows make excellent mothers, farrowing large litters of 10–15 piglets. They have plenty of milk.

Kune Kune

docile • comes in various colours • intelligent • ideal pet

TYPE: stocky breed, with short legs and a hairy body • SIZE: small • COLOUR: various, from gold, ginger and brown to black and spotted combinations

The Kune Kune (pronounced 'coo-nee, coo-nee') comes from New Zealand, although the breed is not indigenous to that country. No one is really sure where the Kune Kune originally hails from and there are several theories about how the pigs came to New Zealand. The evidence of pigs still surviving on the South Pacific islands, which have similar distinctive tassels to those of the Kune Kune, suggests that the breed may have been taken from there to New Zealand as a meat source by the Maoris. New Zealand's early settlers may also have transported Kune Kune, or it is possible that they were released in New Zealand by whalers and sealers to provide them with easily sourced food on future voyages.

Kune Kune

Once sought-after by the Maoris for food, the Kune Kune breed was threatened with extinction during the 1970s as the Maoris' dietary habits changed. Two wildlife park owners, on hearing of the breed's plight, bought up every Kune Kune they could find in New Zealand. They managed to find just 18, but this was enough, with later additions, to form a stud book.

Today the population of Kune Kunes is in a much better state, with people all over the world wanting to buy them. Kune Kune piglets are now in great demand and fetching high prices.

Kune Kunes arrived in Britain in 1992 when it was decided, in case disease occurred in their home country, that another population should be set up elsewhere. As many variations as possible of genetic stock were brought into the country, to ensure that a true representation of the breed would be preserved. The Maoris traditionally preferred black pigs, but today Kune Kunes come in many different colour variations. In 1993 and 1996 further bloodlines were imported into the UK, firmly establishing this breed.

Looks

The Kune Kune is covered in hair, which can be short and straight or long and curly. Hair colours include cream, gold, ginger, brown, black and spotted combinations. It has a medium to short snout, and either pricked or semi-lopped ears. It has short legs and a short, round body – in fact, the name Kune Kune means 'round and fat'. An unusual feature of this breed are the two tassels called *piri piri* under the

This small breed is perfect for keeping in large gardens.

Kune Kunes love humans and make delightful pets.

chin, which are similar to those of a goat. The Kune Kune stands about 60 cm (24 inches) tall.

Personality

The Kune Kune is a delightful pig and suitable for the novice owner. It is placid and friendly and loves human company. The breed is intelligent and easy to train, as many owners of these animals will readily confirm.

Habitat

The natural habitat of the Kune Kune is woodland and pasture. They love being outdoors and are suitable for a range of climates. Although this small pig is potentially ideal for a garden, owners must ensure that it is given adequate space as the Kune Kune is happiest when it has a big area to wander about in.

Management

The Kune Kune is easy to manage, as it has the ability to fatten on little more than grass, as well as a disinclination to roam. It does not do well on too much protein and needs more fibre than most pigs, so its feed should contain no more than 16 per cent protein. To mature to a decent-sized cut, this breed needs to be kept for longer than most other types of pig.

Breeding

Kune Kune boars become fertile at six to seven months and the gilts can get pregnant as early as five months. You should, however, wait until the gilt is at least 12 months old before trying to mate her. Kune Kune sows make good mothers, but the litters vary considerably in size.

A Kune Kune needs a plentiful supply of water, but will gain weight on little more than grass.

Vietnamese Potbelly

excitable • easy to train • firm, prickly hair • good pet

TYPE: small breed with a sway back and pot belly • SIZE: small • COLOUR: wide variety of colours, such as black, silver, white or tan

Vietnamese Potbelly pigs were first developed in the 1960s, originally from the Mong Cai pig, which was once found right across Vietnam. They were initially exported to Sweden and Canada. In 1986, the first pigs were sold in America, achieving a price of thousands of dollars. When the breed was first developed, the colour of the Potbelly was predominantly black, with the odd spot of white here and there, but today the pigs come in a wide range of colours.

Due to cross-breeding in its native country, the indigenous Vietnamese pig species is not as prolific as it once was and exists only in mountainous Vietnam and Thailand. The Vietnamese government has therefore begun

Vietnamese Pot Belly

to subsidize local farmers who continue to raise Potbelly pigs. A breeding programme is in place to ensure that the breed will not die out completely.

The most famous Potbelly pig was Max, belonging to the actor George Clooney, who reportedly said of his 18-year ownership of his pet that it was the longest relationship he had ever had.

Looks

The Vietnamese Potbelly is a relatively short animal, reaching only 40–50 cm (16–20 inches) tall when it achieves maturity. It has short legs, a straight tail and a large pot belly. The hair is different from that of other pigs, being prickly and firm.

Personality

Although highly intelligent, the Vietnamese Potbelly is known for being aggressive and excitable. Its temperamental character means that it is not ideal for beginners. It can easily become overweight.

Habitat

This breed should be allowed access to its natural environment of grass or woodland.

Management

Most people keep the Vietnamese Potbelly as a pet, but it still needs to be treated as a normal pig. It should be allowed to graze and root, and should not be overfed. In the wild this breed eats grubs, berries or anything else the pigs can get hold of. Proper pig food with less than 16 per cent protein should be given to them, along with vegetables and fruit.

Like any other pig, the Vietnamese Pot Belly needs the space to graze and root.

Breeding

The average litter size is six to eight piglets, but it is not unknown for 12–15 to be born weighing less than 0.45 kg (1 lb) at birth. Their lifespan is usually 10–15 years, although some may live longer.

Meishan

slow-growing • wrinkled face and skin • disease-resistant • succulent meat

TYPE: strong, fat breed with lop ears • SIZE: small • COLOUR: black or grey

Meishan (pronounced 'may shawn') pigs were exported from China to the United States in July 1989. Originally they came from the lowland areas of China and were considered to be Taihu pigs – taking their name from Lake Taihu in the Lower Changjiang River basin of north-central China.

Looks

These smallish pigs have a large head, a wrinkled forehead and lop ears. During pregnancy and lactation the sow's belly is so low that it nearly touches the ground. The thickness of their back fat is approximately 2.5 cm (1 inch). Their coat is black and sparse.

Personality

The Meishan is a docile breed.

Habitat

This pig does best in a free-range, outdoor environment.

Management

Meishan pigs are slow-growing and reasonably fat, compared to modern commercial breeds. It is thought that this breed of pig is particularly resistant to disease. In their native China the Meishan's diet consists of barley and rice bran, as well as large amounts of roughage and plants that live by water.

Breeding

Meishan pigs are efficient reproducers, usually having large litters containing 12 piglets in the first litter, followed by as many as 15 in consecutive litters.

Meishan

The Meishan is a lard pig, in contrast to the leaner pork breeds now popular in the West.

Questions and answers

Don't be put off keeping pigs because you're not sure about some aspect of their care or welfare. In this section you'll find answers to many of the commonly asked questions, cross-referenced to more detailed explanations elsewhere in the book.

Are pigs difficult to keep?
If you are keeping pigs solely for meat, then they are one of the easiest animals to look after. They are hardy and, as long as their basic welfare needs are met – in other words, decent shelter, the correct amount of food and a sufficiently large area to wander about in – the amount of work required is minimal. Visiting your pigs once or twice a day to check that they are well, and to feed and water them, is usually all that is needed.

Which breed is best for beginners?
Pigs with lop ears (ears that cover the eyes) are usually considered the ones most suitable for beginners as they tend to have a more docile nature than pigs with prick ears. Breeds such as the Large Black (see pages 134–137) and the British Saddleback (see pages 128–131) are just two breeds with lop ears that are ideal for novice pig-keepers, and both of these breeds may be reared for their pork and their bacon.

Will my garden suffice to keep two weaners?
When choosing a breed, you need to take into account the area of land that you will be putting your pigs on. If you are planning to keep them in a garden – which is certainly feasible – your choice of breed is limited to a small or medium-sized pig, such as the Kune Kune (see

Breeds with lop ears, such as the Gloucester Old Spot, often have a gentle nature and are suitable pigs for beginners.

Pigs love to root and will dig up your garden in no time at all if you let them.

pages 184–187) or the Berkshire (see pages 132–133). However, if you are in the fortunate position of owning a few acres, you can choose whichever breed takes your fancy. So long as you have the minimum amount of land required to keep a pig (see page 42), you should be able to find a breed to fit your space.

Do I need to check my pigs every day or will once or twice a week do?

Pigs must be checked at least once a day – and preferably twice – to ensure that they are in good health, have sufficient water (especially in hot weather) and to feed them. Even if your pigs are kept in woods that are full of nuts and berries and other types of food, or in a field full of lush grass so that they gain most of their food intake by foraging, they must be visited daily.

Will pigs ruin my land?

The short answer is: yes, they will ruin the land, if you let them. If you are concerned that this might be in issue, get a breed that is less likely to root, such as the Middle White (see pages 144–145). Ideally, pigs should be rotated regularly on different areas of land, and given as much land as possible to wander around in, to rest the ground and prevent it becoming 'pig sick'. One solution is to divide your ground in two, if you have sufficient space, and keep your pigs on one half while resting the other half.

Where will I find pigs for sale?

The best sources of quality stock for sale are the breed societies. Most of their websites have a 'stock for sale' page and this is a good place to start your search. Specialist smallholder magazines and newspapers also usually advertise a selection of breeders. No matter where you buy from, before parting with your money make sure that you have seen the piglets and satisfied yourself that they are healthy.

At what age should I buy weaners?

Generally, weaners are piglets that have come off the sow at between eight to ten weeks. Most breeders of traditional outdoor pigs wean at eight weeks. If you buy a weaner older than this – even by just a week or so – you should be prepared to pay more than you would for an eight week old.

Is a male or female pig best for pork?

There is no difference in the taste of pork from a male or a female pig killed at six months, but most butchers will not buy the carcass of a male pig due to the possibility of 'boar taint', which is the term for the strong, unpleasant smell and flavour that the meat takes on due to the pig reaching sexual maturity. However, boar taint is not usually found in traditional breeds – even in older boars – that are kept outside; it is more often associated with intensively reared animals, and is believed by some people to be attributable to stress in the boar.

Will a pig make a good pet?

Some pigs, especially the bigger breeds, are not ideal for keeping as pets, because they can grow into large, unruly animals, which – if not managed correctly – can cause a great deal of damage. The smaller breeds adapt better to a pet's lifestyle than the larger pigs and can be kept in a more confined area. However, although pigs are intelligent animals, you should not expect them to behave as a dog would with its owner. Pigs should always be kept in pairs because they are sociable animals, and keeping one pig on its own amounts to cruelty. And every pig (whether or not it is a pet) should have access to outside space of an adequate size.

Aren't pigs dirty?

Pigs are definitely not dirty animals – they will toilet in one corner of their territory, usually the one furthest away from their sleeping area. Having said that, occasionally a pig will dirty its pen, but this is rare (see page 62). Pigs shouldn't smell, either, if you only have a couple of animals, although if you have male pigs, you may find they do smell.

What do I need in the way of licences in order to keep pigs?

In the UK, by law you need a holding number (CPH number, see page 44) in advance of getting your pigs; and, once you have the animals, you need to acquire a herd number, which is the number given to you as a pig owner and accompanies the pigs wherever they go (even if you buy in some new pigs, they should be given the same herd identification number). Pigs can be identified with this number in one of three ways: by slapmarking, tattooing or ear-tagging (see pages 25–26). If you are unsure what regulations apply in your own country, contact the relevant government department or animal-welfare organization for advice.

Weaners are young pigs, between eight and ten weeks old, that are no longer suckling from the sow.

Small docile breeds, such as the Kune Kune, are ideal for families with children.

What records do I need to keep, once I have my pigs?

You will need to keep (either on paper or on your computer) your pigs' movement records and medical records. And once a year in many countries you will need to record how many pigs you keep under your holding number and send this information to the relevant authority. See pages 44–45 and check with the relevant authority for further information.

Can I leave my pigs when I go on holiday?

As with any animal you keep, whether it is a dog or a fish, you have a responsibility to make sure that it is checked, fed and watered at least once a day if you go away. However, if pigs are kept in a large area where they can source a certain amount of food (such as berries) for themselves, it is possible to leave them for a day and a night, as long as they were fed before you left and are fed and checked again as soon as you come back the next morning. It may be worth trying to befriend other local smallholders, so that you can take it in turns to look after each other's livestock when you go away for longer periods.

Is it safe to allow pigs near children?

Depending on the breed and, of course, on the nature of your own animals, children can be encouraged to mix freely with your pigs. However, never allow a child in with a boar, no matter how friendly that boar is. If you have a young family, opt for a docile breed such as a Gloucester Old Spot (see pages 126–127).

Can I feed kitchen scraps to pigs?

In some countries it is illegal to feed kitchen waste to pigs (in the UK this law was brought in after the foot-and-mouth outbreak in 2001). If that is the case you will only be allowed to feed products such as vegetables that have come straight from your garden or from a place that deals only in vegetables. Under no circumstances will you be allowed to feed meat or meat by-products. In any case, it may be best to avoid feeding your pigs kitchen scraps due to the possibility of spreading viruses and bacteria and to the likelihood of your pigs putting on too much weight.

I live in a cold part of the country. Does that matter when keeping pigs?

Traditional breeds are renowned for their hardiness and their ability to live in any climatic conditions. But some breeds are not as hardy as others, and if you live in a particularly cold area, you should avoid buying white-skinned pigs. Instead, choose a coloured pig such as a Large Black (see pages 134–137) or one with a hairy coat such as the Tamworth (see pages 138–141).

The hairy coat of the Tamworth keeps it warm in cold weather, while its reddish pigmentation protects it from sunburn.

I will be keeping pigs next to my horses. Will this cause any problems?

Your horses will be fine, once they get used to the pigs. Riding a horse that is unfamiliar with pigs close to a field that contains them may lead to difficulties. However, if your horses live in a field adjacent to the pigs, in no time at all they will be touching noses over the fence.

I keep free-range chickens. Will the pigs attack them, if the chickens stray into the pig pen?

There have been a few incidents in which pigs have killed and eaten chickens, but there are also many other situations in which chickens live quite happily in the same field as pigs. However, problems may arise if the chickens find themselves shut in a building with pigs or enclosed with them at feeding time (when the pigs will be protective about their food). If possible, try to feed the pigs well away from the chickens, to prevent your birds becoming an extra source of food at mealtimes.

I would like to have a go at breeding, but don't want to buy a boar. How can I get my gilt in-pig?

If you don't want to buy a boar, there are companies that specialize in providing semen from pure-bred boars, which enables you to artificially inseminate your sows. However, if this idea does not appeal, see if you can find a boar locally to which you can take your sow (see page 87). Breed societies often advertise boars that are at stud, so ask around to find out where your nearest boar is.

I am already getting attached to my pigs. How will I cope when the time comes to have them slaughtered?

The first time you take any animal to slaughter is always difficult. Not naming your pigs may help you stay detached from them. But if you feel you will be unable to take them to the abattoir yourself, maybe you can enlist someone to take them for you. Remember that your pigs have had a good life. If we don't eat the pigs we keep, the traditional breeds will die out.

Novice breeders may want to buy a gilt that is already in-pig or, alternatively, try using artificial insemination.

In the course of caring for their animals, owners often develop a strong bond with their pigs.

Pigs are very sociable animals and they should not be kept on their own.

Where we live straw is difficult to source cheaply, if indeed we can find a farmer to sell us some. What's the alternative?

You don't always need to use straw to bed down your pigs. In fact, when it's hot in summer they often prefer to sleep on the cool earth. If straw is difficult to come by, woodshavings are a good substitute, although they can be expensive and they also have a tendency to be slippery underfoot. If you live near a forest, why don't you try gathering bracken and drying it to use? Alternatively, long grasses such as reeds also work well. Some pig owners have even used a pile of dried leaves in the ark, much to their playful pigs' delight. Another option is buying horse mats (used to bed down horses). These mats are extremely easy to clean and very durable, but they can be difficult to manoeuvre into the ark. (See pages 22–23 for more on bedding for pigs.)

How can I tell if my pigs are too fat?

Use this rule-of-thumb method to tell if your pig is overweight: press down on its back and if it is the ideal weight you should be able to just feel its backbone. If you can't, your pig probably has too much backfat. If you can see the backbone then the pigs are too thin. Female pigs tend to put on fat quicker than the males once they get to a certain age. However, if your pigs are living a healthy outdoor life and you are not overfeeding them, then they shouldn't get too fat (see pages 52–55).

Can I keep a pig on its own?

Keeping a pig on its own for any length of time is cruel. They are very social animals and they need company of their own sort, even if there is another animal, such as a goat, already sharing their paddock. Pigs can get very depressed if they do not have another pig of around the same age to play with.

My budget does not run to refencing the area where I would like to keep pigs. Will electric fencing on its own keep them in?

It is certainly possible to keep whole herds of pigs behind electric fencing. However, unless they have been brought up with electric fencing, this alone will not keep them in. Unlike other animals who back off when they encounter electric fencing, pigs tend to rush through it. A back-up fence is strongly recommended in case they do break through (see pages 29–31).

My pigs have dry skin. What should I do?

Once you have checked that the dry skin is not caused by lice or mange (see pages 72–73), buy some baby oil and rub that well in until your pig's skin starts to gain moisture. If your pig will let you, give him an exfoliation by brushing the skin with a horse brush. Then, once you have brushed off all the loose dried skin, rub in the oil well. You may have to do this three or four times a week until you see some improvement.

My pigs seem to be permanently wading through mud. Will this damage their legs and feet?

Pigs like to trample and root and, if it rains a lot, it is impossible to keep your pig's enclosure from getting muddy. Although it isn't pleasant for the pigs to be continually plodding through mud, it won't harm them if they can go to a dry area, such as an ark with a floor. Keep plenty of straw in the ark to help dry the mud on the pig's legs. If possible, try to move the troughs around, to avoid any one area becoming particularly bad. Never distribute feed on the ground if it is muddy, as the food will just disappear into the mud – you will be wasting your money and the pigs won't get properly fed. If you have enough space, it is a good idea to rotate the land use, to allow the land to recover after it has been used as a pig enclosure.

Glossary

Ad lib feeder Feeder for piglets and weaners that allows them to take food in small quantities as needed

AI Artificial insemination

Ark Moveable outdoor pig housing with curved roof

Back fat The depth of fat along the pig's back

Bloodline The ancestry of a pure-bred pig

Boar Uncastrated male pig

Boarlet The young of the wild boar

Carcass Dressed body of the pig

Creep A partitioned-off area with a heat lamp where the piglets can sleep safely

Creep food High-protein pellets given to piglets from three weeks

Colostrum First milk after farrowing, full of antibodies

Dished With an upturned snout

Farrowing Giving birth

Finish Last stages of fattening

Foundation stock Breeding stock

Gestation The period of pregnancy, which in pigs is 116 days

Gilt Female pig that has not yet produced a litter

Grazer Pig that prefers to graze, rather than root

Grower food High-protein food that is given from two weeks before weaning

Ham Meat taken from the back leg of the pig; this part of the pig's body

In-pig Pregnant

Lactation Period when the sow is producing milk

Lop-eared Pig with ears that hang forwards over the pig's eyes

Meal Flakes mixed with water to form a mush

Nipple drinker Watering system that is activated by the pig moving the drinker

Paddle Combined slapstick and pig stick, used for moving pigs

Pig nuts Large nuts fed to dry sows and boars

Pig stick Handling stick, used in conjunction with a slapboard for moving pigs

Porker Pig fattened for food, at around five to six months

Prick-eared Pig with ears that stand upright

Rooter Pig that has a tendency to root, often with a long snout

Scours/Scouring Diarrhoea

Slapboard Board with a hand-hole used in conjunction with a pig stick for moving pigs

Sow Female pig after farrowing the first litter

Sow nuts Nuts fed to weaners, in-gilt pigs and suckling sows

Store pig Fattening pig between weaning and slaughter age

Terminal sire Breed of boar, such as the Duroc, used extensively in improving other breeds.

Underline Teats underneath a pig

Wallow Hole containing mud and water where the pig can cool down during hot weather

Weaner Piglet that is no longer suckling from the sow, usually 8–10 weeks

Weaner nuts *See* sow nuts

Withdrawal period Time during which an animal may not be used for food production following administration of a medicine

Index

Page numbers in *italic* refer to the illustrations.

A
abattoirs
  butchers at 103
  collecting carcass from 102, 105
  and garden pig-keeping 43
  and 'liver spot' 71
  marking pigs 24, 25, *25*
  paperwork 101
  slaughter process 102
  taking pigs to 60, 98–9, *99*, 198
acorns *75*
acquiring stock 36–9
ad lib feeders 53, *53*
advertising 94–5, 112, 194
African swine fever *77*
afterbirth 62, 83, 89
agalactia *76*
agility classes 111, *114*
agricultural shows *see* shows
aloe vera 69, *76*
American Landrace 120, *123*
American Yorkshire 166–7, *166–7*
anaemia 73, 74
Animal Health Offices 44, 76
animal welfare 42–3, 98–9, 101
anthelmintic preparations 28, 71
anthrax *77*
anti-bacterial handwash 62, 65
antibiotics 28, 48, 69, *77*
antibodies 90
anti-parasitic solutions 73
antiseptic sprays 28
appetite, loss of 68
arks 20, *20*, 21, 32–3, *33*
  bedding 40, *41*
  cleaning 21, 51, 62, *64*
  dunging in 62
  farrowing arks 62–4, 82, 88, 90
  farrowing crèches 82–3, 88, 90
  floors 21, 33, 62, *63*
  hygiene 62
  keeping warm 27
  siting 33, 40–1
  sizes 32
artificial insemination (AI) 86–7, *86*, *87*, 198
Asian pigs 11
atrophic rhinitis 74, *74*
auctions 36, *94*

B
baby oil 28, 201
backfat *55*, 201
bacon pigs (baconers) 12, 16, 100
bacteria *54*, 71–2, 197
balls, playing with 35
bananas 68
barbed wire 31
barley meal *55*
barns 34–5, 40, 62, 82
bedding 22–3, *23*, 40, *41*, 201
  in farrowing sheds 88
  hygiene 51, 62
  in isolation pens 65
  powder disinfectants 65
Bentheim Black Pied 160, *160–1*
Berkshire 11, 12, 132–3, *132–3*
  garden pig-keeping *42*, 43, 193
  meat 16
bio-security 62–5, *77*, 114
birds 21, 65, 72
birth 88–9
black pigs, oiling *111*
bloodlines 80, 87, 108
boars
  'boar taint' 98, 194
  bringing to sow 87
  buying 86
  feeding quantities 54
  getting sows in-pig 84–7, *85*
  handling 61
  keeping your own 84–6
  meat from 98, 194
  safety 61, 197
  selecting for breeding 80–1, *80*
  smell 86, 194
  stud boars 84
  taking sow to 87, 198
  teats 81
  temperament 38, *80*, 81
  tusks 61, 86
  water requirements *57*
boots 26, *26*, 62, 65
boredom 50
bracken *75*, 201
breed societies 15, 36, 45, 94, 115, 194
breeders
  at agricultural shows 16, 81
  buying pigs from 36–8
  getting to know 109, 115
  showing pigs 108–9
breeding 17, *79–95*
  buying pigs for 38–9
  farrowing 88–9
  getting sows in-pig 84–7, *85*, 198
  marketing piglets 94–5, *94*
  records *45*
  selecting boars and sows 80–1
  temperament and 38
breeds 11–12, 116–91
  bloodlines 80, 87, 108
  breed standards 111
  choosing 14–17, 94, 192
  for garden pig-keeping 43
brick walls 32
bristles, removing 102
British Landrace 12, 16, 98, 120–2, *120–1*
British Lop 148, *148–9*
British Pig Executive (BPEX) 101
British Saddleback 12, 128–30, *128–31*, 192
  choosing 39
  cross-breeds 15
  identifying *25*
  markings 81
  meat 98
brooms 26
brushes 27, 201
buckets 20, 21, 23, *23*, 58, 65, 83
budgeting 8–9, 20–1
buildings 32, 34, *34*, 40
  see also arks
bulk buying 52
bullying 48
butchery 103–4
buying
  boars 86
  equipment 20
  food *52*
  pigs 8, 36–9, *36*, 194
by-products 12

C
calamine lotion 28, 71
calcium *55*
cameras, surveillance systems 83, 88
carcasses
  butchery 103–4
  disposing of 69
catching piglets 58–9
cave paintings 10
Certificate of Competence 101
Chester White 178–9, *178–9*
chickens 198
children 111, 197
classical swine fever *77*
cleaning

arks 21, 62
equipment 26
troughs 51
clearing land 9
clipping teeth 83, 90
clothes 26, *26*
clubs 115
cod-liver oil 55
cold weather 54, 197
colds 73
colostrum 72, 90, 92
compound food 53–4
conformation
breed standards 111
buying breeding stock 81
faults 108
conservation, traditional breeds 37
containers
feeding 22, *22*, *52*
storing food 23, *23*, 65
water 21
contamination, preventing 105
cooperatives 9
costs 8–9, 20–1
cottagers' pigs 11
cotton wool 28
coughing 38, 73
County Agricultural Offices, Northern Ireland 44
CPH numbers 44, 194
crackling 98
crates, farrowing 82
crèches, farrowing 82–3, 88, 90
creep areas 83, *83*, 88, 90
creep food 53, 54, 92
creep lamps 83
cross-breeds 14–15, 37
cupboards, medical 28
customers, showing pigs and 109
cuts and grazes 48
cuts of pork 104, *104*
cutters 100

D

daily tasks 48–50
dead pigs, disposing of 69
deadly nightshade 75
Defra 76
dehydration 72
Demodex phylloides mites 73
depression 37, 72, 201
diabetes 12
diarrhoea 53, 72
disease 17, 67–77
bird droppings and 65
exemptions for registered stock 37
isolation pens 65
kitchen waste and 54, 197
notifiable diseases 76–7
preventing 62
stress and 40
traceability and 44
disinfectants 27, 51, 62, 64–5, 102
documentation 27, 40, 44–5, 50, 197
dogs 65
domestication 10
doors
farrowing arks 88
pigsties 34
Dorset Gold Tip 11
drainage, pigsties 34
drinkers
automatic drinkers 20, 21, *21*, 65
cleaning 49, 51, 65
nipple drinkers 21, *56*, *57*
drugs *see* medicines
dry skin 28, 201
dunging areas 62, 194
Duroc 37, 170–2, *170–3*

E

E. coli 71–2, 75
ears
lop-eared pigs 15–16, 39, 58, 192
prick-eared pigs 11, 16, 58, 192
notching 25, 80
rubbing 58
tagging 24, 26, 39, 44, 80, *102*
tattoos 25, *25*, 44, 80
elder 75
electric fencing 30–1, *31*, 201
checking 48–9, *48*
garden pig-keeping 43
and moving pigs 58
enclosures
choosing pigs 192–3
fencing 29–31, *29*, *31*, 40
garden pig-keeping 42–3, *42*, 192–3
gates 30
inspections 50
moving pigs into 40
'pig sick' land 194
siting arks in 33, 40–1
sizes 42
worms 71
energizers, electric fencing 30–1
entertainment for pigs 35
equipment 20–8, 82–3
erysipelas 65, 72
escaping 29, 40, 43, 60, 93
Essex breeds 12
ethics, showing pigs 112
EU regulations, taking pigs to abattoir 101
exfoliating skin 201
exotic diseases 77
eye-teeth 83, 90

F

farmers' markets 9, 12
farrowing 88–9
arks 82, 88, 90
crates 82
crèches 82–3, 88, 90
equipment 82–3
sheds 62–4, 82, 83
fat content, meat 98, 103
fat levels, pigs 55, 201
feed merchants 52
feeding 52–5
ad lib feeders 53, *53*
boars 84
containers 22, *22*, *52*, 65
garden pig-keeping 43
hand-rearing piglets 92
hygiene 65
kitchen waste 54, 68, 197
loss of appetite 68
new pigs 41
piglets 92
quantities 54
routines 48, 52
storing feed 23, *23*
weaker pigs 48
feet 76, 201
fencing 29–31, *29*, *31*
checking 40, 48–9, *48*
garden pig-keeping 43
see also electric fencing
fevers 69
fighting 48, 84
fines 50
finisher pellets 54
finishing pigs 95
fishmeal 55
flaked maize 55
floors
arks 21, 33, 51, 62, *63*
and lameness 76
pigsties 34
'floppy' pork 103
folklore 11
food *see* feeding; meat
food chain information forms 101
food regulations 105
Food Standards Agency 105
foot-and-mouth 17, 37, 54, 76–7, 197
footballs, playing with 35
footbaths 27, 62, 65
footpaths 30
foraging 10–11, *54*
forests, pannage 10, 11
forks 26
foundation stock 39, 80–1
foxes 83, 90
foxgloves 75
freezers 23, 98, 105
French Landrace *122*
fruit 54, *55*, 68, 72

G
garden pig-keeping 42–3, *42*, 192–3
gastric ulcers 74
gates 30, 43, 60
general-purpose antibiotics 28
Genetically Modified (GM) crops 52
gestation period 88
gilts
  buying 8, 81
  in-pig 39
  mating 84, 86, 198
  teats 81
  water requirements 57
  worming 71
Gloucester Old Spot *13*, 126–7, *126–7*
  cross-breeds 15
  feeding 52
  legends 11
  sausages 104
  temperament 16, *192*, 197
gloves 26, 28, 64
grazing 15, 53
ground, clearing 9
grower food 54, 92

H
Hampshire 168–9, *168–9*
hand-rearing piglets 92
handling pigs 58–61
hands, washing 62, 65
hard water 57
hardiness 197
harnesses 29
hauliers, paperwork 101
health checks 48
health problems 70–7
healthy pigs 68–9
heart valves 12
heat lamps 69, 83, 90
heat mats 90
heatstroke 28, 70, *70*
herd numbers 24, 25, *25*, 40, 44, 194
hereditary faults 81
Hereford 180, *180–1*
hessian 27
hinges, gates 30
history 10–12
holding numbers 40, 44, 194, 197
holiday care 197
home butchering 103
horse brushes 27
horse mats 201
horses 198
hosepipes 26
hot weather 49, 68, 70
housing 21, 32–5, *33–5*
Howitt report 12
hygiene 54, 62–5, 105

I
Iberian 154, *154–5*
identification 24–6, *25*, 44, *102*, 194
illness *see* disease
immunoglobulins 90
in-pig gilts 39, *57*, 71
infections
  colostrum and 90
  hygiene 62
  isolation pens 65
infra-red lamps 83
injections
  anti-parasitic solutions 73
  doing it yourself 69, 77
  iron injections 90
  medical records 45
  vaccinations 51, 69, 72
  worming 69, 71
insemination, artificial 86–7, *86*, *87*, 198
inspections, home butchering 103
insulation 33, 34
insulin 12
intelligence 58
internet 12
  buying equipment 20
  buying pigs 194
  internet forums 115
  selling piglets 94
iodine 83
Iron Age pig 142, *142–3*
iron deficiency 74, 90
isolation pens 65, 114

J
joint ill 74
judges 108, 112–13

K
killing-out percentage 103
killing pigs 102
kitchen waste 54, 68, 197
knives 27
Kune Kune 16, 184–6, *184–7*, *196*
  feeding 53
  garden pig-keeping 15, *15*, 192–3

L
labelling meat 105
laburnum 75
Lacombe 174, *175*
'Ladies' Pig' 16
lameness 76
lamps, heat 69, 83, 90
Landrace 12, 16, 98, 120–2, *120–3*
lard pigs 12
Large Black 17, 134–7, *134–7*
  choosing 39, 192
  hardiness 197
  identifying 25
  temperament 15–16
Large White 12, *12*, 118, *118–19*
laws *see* regulations
lean-meat percentage 103
leaves, as bedding 201
legends 11
legislation *see* regulations
legs, lameness 76
lice 35, 38, 72–3, *73*, *77*, 201
licences 44, 194
  feeding milk and whey to pigs 55, 68
  movement 44–5
limescale 57
Lincolnshire Curly Coat 11, 156
litters
  breeding records 45
  buying weaners for breeding 81
  choosing pigs 37–8
'liver spot' 71
livestock markets 36
loading bars, arks 21, 33
lop-eared pigs 15–16, 39, 58, 192

M
maize, flaked 55
Mangalitza 12, 156–8, *156–9*
mange 38, 73, 83, 201
marketing piglets 94–5, *94*, 108–9
markets 36, *94*
marking paste 24–5
markings, buying breeding stock 81
mastitis 75–6
mating 84
mats
  disinfectant 27, 62
  horse 201
meal 54
measuring a pig 100, *100*
meat 11
  breeds of pig for 16, 98
  butchery 103–4
  buying breeding stock 81
  categories 100
  crackling 98
  cross-breeds 14–15
  cuts 104, *104*
  fat content 98, 103
  feeding to pigs 197
  gender differences 194
  labelling 105
  lean-meat percentage 103
  meat-marketing schemes 37
  selling 9, 105
  shelf life 105
medical uses of pigs 12
medicines
  medical equipment 28
  natural medicines 76
  records 45
  withdrawal periods 69, 99
Meishan 190, *190–1*

meningitis 72
metal tags 24, 26, 44, *102*
'Mexican hat' cast-iron feeders 22, *22*
mice 65
Middle White 12, 144–5, *144–5*
  grazing 15, 53, 193
  identifying 25
milk
  agalactia 76
  dried milk 92
  feeding pigs 55, 68
  farrowing 88, *88*, 90
minerals, feeding pigs 55
mismarked pigs 81
mites 35, 73, *73*
molasses 54
monthly tasks 51
movement licences 44–5
moving pigs 40, 58–60, *59*, *60*
mucking out 34–5, 51
mud 201
  wallows 28, 41, 70, 71, *71*
muscle injections 69

N
naming pigs 58, 99, 198
national requirements, record-keeping 44
natural medicines 76
navel 74, 83
needles, injections 28, 69
neighbours, and garden pig-keeping 43
nesting 88, *89*
netting 30, 31
networking 109
nightshade, deadly 75
nipple drinkers 21, *56*, *57*
Northern Ireland 44
nose rings 27
notching ears 25, 80
notebooks 27
notifiable diseases 76–7
nuts 22, 54, *54*

O
oiling pigs 28, 111, *111*, 201
organic pig-keeping 52, 69
over-trousers 26
overalls 26, *26*, 65
overweight pigs 201
Oxford Sandy and Black 17, 124–5, *124–5*
  feeding 52
  garden pig-keeping 43
  piglets *93*
  temperament 16
oxytocin 76, 89

P
paddles 24
painkillers 74
pancreas 12
pannage 10, 11
paper, shredded 23
paperwork 101
parasites 28
paths 30
pedigree pigs
  breed clubs 115
  breeding 17, 80
  choosing 14, 15
  record-keeping 45
  traceability 37
  weaners 8
pellets, finisher 54
penicillin 89
pens
  hygiene 62–4
  isolation pens 65, 114
  walkways 30
pests and diseases *see* disease
pets, pigs as 194
phosphorus 55
Pietrain 150, *150–1*
pig netting 30, 31
pig nuts 22, 54, *54*
pig oil 28
'pig sick' land 194
pig sticks 24, *24*
pig weighing bands 27
pig welfare 98–9, 101
pigboards 24, *24*
piglets
  bedding 22, 23
  breeding records 45
  buying 36–8
  clipping teeth 83, 90
  creep areas 88, 90
  E. coli 71–2
  farrowing 82–3, *83*, 88–9
  feeding 53, 54, 92
  first few days 90
  hand-rearing 92
  iron deficiency 74
  joint ill 74
  marketing 94–5, *94*
  moving 58–9
  piglet booster 83, 90
  preventing sow from lying on 22, 88, 90
  registering 95
  scours 72
  suckling 90, *91*
  water 57
  weaning 54, 92–3
  worms 71, 92
  see also weaners
pigsties 34, *34*
plants, poisonous 40, 50, 75
plastic arks 33, 62
plastic sheeting 27
pliers, tattoo 25, *25*
pneumonia 40, 73
pocket knives 27
poisonous plants 40, 50, 75
Poland China 176, *176*
pork *see* meat
pork pigs (porkers) 12, 100
post-and-rail fencing *29*, 30
potatoes 54
powder disinfectants 65
predators 90
pregnancy 88
pricing weaners 95
prick-eared pigs 11, 16, 58, 192
processing 97–105
progressive atrophic rhinitis (PAR) 74
protein 53, 54
public footpaths 30
pure-bred pigs *see* pedigree pigs

R
ragwort 40, 75
rails, farrowing 82
Rare Breeds Survival Trust 12, 37
record-keeping 27, 40, 44–5, 50, 197
Red Wattle 182–3, *182–3*
reeds, as bedding 201
refrigerated vans 105
refrigerators, storing meat 105
registration 44–5, 194
  buying registered pigs 37, 39, 80
  registering piglets 95
regulations 9, 194
  food regulations 105
  home butchering 103
  showing pigs 112–14
  taking pigs to abattoir 101
respiratory diseases 74
rhinitis, atrophic 74, *74*
rhododendron 75
rings, nose 27
rodents 62, 65, 72
rooting 15, 27, 53, 193, *193*
rope twitches 28
routines 48–51
rubber boots 26, *26*, 62, 65
rubber gloves 28
rules *see* regulations
runts 37, 81, 90
Rural Land Registry (RLR) 44
Rural Payments Agency (RPA) 44

S
Saddleback *see* British Saddleback
safety
  children 197
  electric fencing 30
  handling boars 61
Sarcoptes scabiei mites 73, *73*
sausage pigs 100
sausages 104, 105
scoopers, long-handled 26

Scotland 12
Scottish Executive Environment and Rural Affairs Department 44
scours 53, *72*, 93
scratching 58
season 84–7, 93
sedatives 89
selling
  meat 9, 105
  showing pigs and 108–9
  weaners 94–5
semen, artificial insemination 86–7, *86*, *87*, 198
service, keeping your own boar 84–6
shade 33, 71
shavings, bedding 22–3, 88, 201
sheds, farrowing 82, 83
sheeting, plastic 27
shelters 32–5, *33–5*
  daily tasks 48
  garden pig-keeping 43
  hygiene 51, 62
  straw 20, 21, 51
shoulder harnesses 29
shovels 26
showing pigs 107–15
  choosing and buying pigs 16, 36, 81
  moving pigs 60
  preparation 110–11, *110*, *111*
  rules 112–14
  selecting pigs for 110–11
shredded paper, bedding 23
sick pigs 54, 65
  see also disease
silos 52
skids, arks 21, 33
skin
  brushing 27
  choosing pigs 38
  crackling 98
  dry skin 28, 201
  injections 69
  lice 72–3, *73*, *77*
  mange 73
  scratching 58
  sunburn 28, 49, 71
slapboards 24, *24*, 59, *59*, 60, 61, 112
slapmarks 24, 25, *25*, 44, *44*
slaughter process 102
  see also abattoirs
smallholders' clubs 115
smells, boars 86, 194
snout
  atrophic rhinitis 74, *74*
  nose rings 27
  rooting 15
  snub noses 15
  twitches 28
sociability 201
sodium 55
sow nuts 54
sows
  agalactia 76
  artificial insemination 86–7, *86*, 198
  breeding records 45
  bringing boar to 87
  farrowing 88–9
  farrowing equipment 82–3
  fat levels 55
  feeding quantities 54
  getting in-pig 84–7, *85*
  mastitis 75–6
  meat 194
  moving piglets 58
  preventing from lying on piglets 22, 88, 90
  season 84–7, 93
  selecting for breeding 80–1
  shoulder harnesses 29
  taking to boar 84, 87, 198
  temperament 38
  water requirements 57
  weaning piglets 92
  worming 71
soya meal 55
spades 83
Spots 177, *177*
squealing 59
standstill, 21-day 45, 65, 114
sticks 24, *24*, 59, *59*, 60
stillbirth 88
stock netting 30
stocking ratio 30
storage
  feed 23, *23*
  identification equipment 26
  meat 98, 105
  medical equipment 28
store pigs 34–5, 60, 61
straining posts 30
straining wire 30
straw
  bedding 22, *23*, 40, *41*, 51, 201
  nesting 88, *89*
  shelters 20, 21, 35, *35*, 51
stress 58
  and farrowing 88
  gastric ulcers 74
  moving pigs 40
  taking pigs to abattoir 98
stud boars 84
subcutaneous injections 69
suckling 90, *91*
suckling pig 100
sunburn 28, 49, 71
suncream 28, 71
superstitions 11
surveillance systems, farrowing sheds 83, 88
Swabian-Hall Swine 152–3, *152–3*
swine fever 77
swine pox 75
swineherds 10
syringes 28

T

tags, ear 24, 26, 39, 44, 80, *102*
Tamworth 11, 12, 138–40, *138–41*
  bacon 16
  choosing 39
  cross-breeds 15, 37
  garden pig-keeping 15, 43
  handling 58
  hardiness 197, *197*
  rooting *14*, 15, 53
tape measures 100, *100*
tattoos 25, *25*, 44, 80
tea-tree oil 76
teats 81, *81*, 90
teeth clippers 83
temperament 16, 38, *39*, *80*, 81, 192
temperature
  in arks 33
  heatstroke 70
  taking 68–9
temporary marking paste 24–5
thermometers 28, 68–9
torches 27
towns, garden pig-keeping 43
traceability 37, 44
Trading Standards Departments 45
traditional breeds 11–12, *17*
  'boar taint' 194
  breed clubs 115
  breeding 17
  conservation 37
  farrowing equipment 82
  fat 98
  hardiness 197
  meat 98
  sunburn 71
trailers 27, 43
  Certificate of Competence 101
  disinfecting 62
  at shows 114
  and stress 40
  taking pigs to abattoir 98–9, *99*, 101
  trailer-training 60–1
trees
  arks under 33
  damage by pigs 50–1
  as shelter 32
troughs
  cleaning 51
  feeding from 22, *22*
  in isolation pens 65
  water 21, 56–7, *57*
trousers 26
tusks 61, 86
21-day standstill 45, 65, 114
twitches 28, 69
tyres 82

U
udders 88–9, *88*, 90
mastitis *75–6*
ulcers, gastric 74
umbilical cord 83
underlines, buying breeding stock 81, *81*
urban areas, garden pig-keeping 43

V
vaccinations 51, 69, 72
vans, refrigerated 105
vegetables 54, 68, 72, 197
ventilation, housing 32, 33
vets 69, 76
inspection at abattoir 102
in urban areas 43
when to call 77
Vietnamese Potbelly 16, 188–9, *188–9*
vinegar 28, 70
viruses 54, 197
visitors, bio-security 64, 65
vulva 88

W
waiting lists 37, 95
walkways 30, 59–60
wallows 28, 41, 49, 70, 71, *71*
washing pigs *110*, 111
water *56–7*, *56–7*
checking 48, 49
daily requirements 57
for new pigs 41
wallows 28, 41, 49, 70, 71, *71*
water butts 21
see also drinkers
weaners 8
breeding stock 39, 80–1
buying 36–7, *36*, 80–1, 194
cross-breeds 15
dunging 62
escaping 93
feeding quantities 54
selling 94–5
trailer-training 60
water *57*
weight 100
weaners' nuts 54
weaning piglets 54, 92–3
weather 49, 54, 68, 70, 197
websites *see* internet
weekly tasks 50–1
weighing bands 27, 100, *100*
weight gain 33, 35, 52, 53, 90
weights, sending pigs to abattoir 100
Welsh Assembly, Divisional Office 44
Welsh pig 12, 25, 146–7, *146–7*
Wessex breeds 12
wheat meal 55
wheelbarrows 26
whey 55
white-skinned pigs, hardiness 197
wild boar 10, *10*, 11, 15, 162–4, *162–5*
windows, pigsties 34
wire twitches 28
withdrawal period, medicines 69, 99
womb 89
wood shavings, bedding 22–3, 88, 201
woodlands 9, 10, 12, 32, 40
Woolly Pig 156–8, *156–9*
worming
frequency 51, 69, 71
medical records 45
piglets 92
wormers 28
worms, symptoms of 38
wound powder 28
wounds 48, *77*

Y
yearly tasks 51

Acknowledgements

Author acknowledgements

In particular I would like to thank Marcus Bates of the British Pig Association, as well as John Wreakes, Brian Card, Bernie Peet, Brian Peacock and the Wales & Border Counties Pig Breeders Association.

Photographic acknowledgements

Special photography © Octopus Publishing Group Ltd/Mick Corrigan

Alamy blickwinkel 152, 153, Gaudencio Garcia 154, Interfoto Pressebildagentur 160, Lynn M Stone/Nature Picture Library 181, Mark Boulton 142, RMT 161; **Corbis** George D Lepp 73 left, Lars Langemeier/AB/zefa 77; **Digital Railroad** Jerome Chabanne 122; **Dr Allison Hughes** MVB CertPM MRCVS 74; **FLPA** Bjorn Ullhagen 86, Sarah Rowland 190, 191; **Getty Images** afp 155, Joel Sartore 182, 183, Raymond Kleboe/Hulton 11; **Grant Heilman Photography Inc** John Colwell 178, 179, Larry Lefever 123; **National Spotted Swine Record, Inc** 177; **Nature Picture Library** Lynn M Stone 166, 167, 180; **PEAK Swine Genetics Inc**175; **Lyndal Rincker** 176; **Science Photo Library** Steve Gschmeissner 73 right.

Publisher acknowledgements

Executive editor Jessica Cowie
Senior editor Fiona Robertson
Executive art editor Leigh Jones
Designer Peter Gerrish
Illustrator Cactus Design & Illustration Ltd
Production manager David Hearn
Picture research Giulia Hetherington and Louise Grimwade